W0228937

11th INTERNATIONAL DOCTORAL STUDENTS WORKSHOP ON LOGISTICS

JUNE 19, 2018

MAGDEBURG

Publisher:
Univ.-Prof. Dr.-Ing. habil. Prof. E. h. Dr. h. c. mult. Michael Schenk

In cooperation with:

With the support of:

This project has received funding from the *European Union´s Horizon 2020 research and innovation programme* under grant agreement No 692426

TABLE OF CONTENTS

Ing. Ernesto González Cabrera
Industrial Engineering Department, Central University from Las Villas, Cuba

DrC. Roberto Cespón Castro
Industrial Engineering Department, Central University from Las Villas, Cuba

Prof. Dr.-Ing. Dr. h.c. Norge Isaias Coello Machado
Mechanical Engineering Department, Central University from Las Villas, Cuba

Dr.-Ing. Dr. h.c. (UCLV) Elke Glistau
Institute of Logistics and Material Handling Systems
Otto von Guericke University Magdeburg, Germany

Olga Morozova
Department of Theoretical Mechanics, Mechanical Engineering and Robotic Systems/Engine Design Faculty
National Aerospace University "Kharkiv Aviation Institute", Ukraine

Vadim Vasilyuk
EOS Data Analytics, Ukraine

Tetiana Pavlenko
Department of Economic Theory/Economics and Management Faculty
National Aerospace University "Kharkiv Aviation Institute", Ukraine

Volodymyr Polovynko
Department of Theoretical Mechanics, Mechanical Engineering and Robotic Systems/Engine Design Faculty
National Aerospace University "Kharkiv Aviation Institute", Ukraine

Tetiana Pavlenko
Department of Economic Theory/Economics and Management Faculty
National Aerospace University "Kharkiv Aviation Institute", Ukraine

Olga Morozova
Department of Theoretical Mechanics, Mechanical Engineering and Robotic Systems/Engine Design Faculty
National Aerospace University "Kharkiv Aviation Institute", Ukraine

Kateryna Pechenizka
Department of Theoretical Mechanics, Mechanical Engineering and Robotic Systems/Engine Design Faculty
National Aerospace University "Kharkiv Aviation Institute", Ukraine

FOREWORD

**Univ.-Prof. Dr.-Ing. habil. Prof. E. h.
Dr. h. c. mult. Michael Schenk**

Managing Director of
Institute of Logistics and Material Handling
Systems, Otto von Guericke University
Magdeburg, Germany

Director of Fraunhofer Institute for Factory
Operation and Automation IFF,
Magdeburg, Germany

Dear Ladies and Gentlemen, Colleagues and Friends,

This year will be held our eleventh International Doctoral Students Workshop on Logistics in Magdeburg. In recent years, the workshop has become a recognized forum for scientific exchange between young researchers from all over the world. Once again, the participants will have the great opportunity to perform to an international audience of scientists and to get in an inspiring exchange with them.

This year we are pleased to welcome participants from Cuba, Hungary, Latvia, Ukraine and several parts of Germany, who will generate interesting insights into their dissertations' topics related to logistics networks and organization, optimization and simulation in logistics as well as on the topic Industrie 4.0 and the meaning of logistics in its context. The presentations and thematic discussion groups traditionally open up a diverse field of the participants' current research findings and give ideas for possible further international collaborations. In addition, this year we will the first time contribute a paper for young scientists, which presents advices and rules for writing a dissertation in an effective and efficient manner.

The workshop is carried out in conjunction with the 21st IFF Science Days, which take place from June 19 to 21, 2018. The various sessions on logistics, digital engineering, plant design and operation and human-robotics collaboration furnish participants an outstanding platform to exchange views on important trends and prospects in their fields, to network and even to initiate new business contacts. This will provide participants an opportunity to benefit from the experience and insights of experts working in both theoretical and practical fields.

The present conference proceedings do not only show the participants' individual research topics, they also furnish insight into the participating international organizations' research and academic work. Pleasingly this year again, our International Doctoral Students Workshop is conducted with the friendly support of the Horizon2020´s project ALLIANCE, whom I would like to expressly thank at this point. I am already looking forward to seeing you at the Twelfth International Doctoral Students Workshop on Logistics 2019 in Magdeburg again.

Sincerely,

Univ.-Prof. Dr.-Ing. habil.
Prof. E. h. Dr. h. c. mult.
Michael Schenk

INTRODUCING
THE UNIVERSITIES

OTTO VON GUERICKE
UNIVERSITY MAGDEBURG

The Otto von Guericke University (OvGU) was founded in 1993 from three institutions of higher education: the Technical University Magdeburg, the Teacher Training College and the Medical Academy of Magdeburg. It is named after the famous scientist Otto von Guericke, whose research on the vacuum, especially his hemispheres experiment, earned him fame beyond German borders.

Consisting of 9 Faculties, OvGU offers more than 70 academic programs. Nearly 14,200 students are enrolled at OvGU; 2,400 of them are international students. OvGU is one of Germany's youngest universities. Its innovative fundamental research contributes to the city's and the country's social and scientific development.

The Institute of Logistics and Material Handling Systems is part of the Faculty of Mechanical Engineering and looks back on more than 50 years of experience in training and research in the field of conveying technologies, logistics and material handling systems.

The fields of research include:

- Mathematical modeling and simulation,
- Development of instruments for analysis and planning,
- The conservation of resources, energy efficiency and sustainable logistics,
- Discrete element method simulation in continuous conveying technology,
- Virtual engineering,
- Ramp-up management and
- The transfer of methodology and know-how in logistics.

www.ilm.ovgu.de

FRAUNHOFER INSTITUTE FOR
FACTORY OPERATION
AND AUTOMATION IFF

The Fraunhofer Institute for Factory Operation and Automation IFF uses custom solutions to help German and foreign companies make their manufacturing smarter. Fraunhofer IFF is a technology partner specialized in planning, developing, equipping and operating work, manufacturing and supply chain systems as well as their supply infrastructures. They design work systems in which humans and machines collaborate side-by-side. They combine these work systems in efficient manufacturing and supply chain systems and use smart infrastructures to connect them with each other and their environment.

Digital engineering integrated throughout product and manufacturing system life cycles is crucial. Fraunhofer IFF achieves this with interoperable methods and tools as well as our expertise in robotics, testing and inspection systems, technology-based assistance and learning systems, and manufacturing and supply chain process engineering.

As a technology partner to companies, they research, develop and improve technologies, systems and products from the idea to manufacturability – and implement them in companies in short time, combining their industry experience and research expertise to do so.

Fraunhofer IFF thus empowers companies to operate adaptively in the marketplace and to boost their manufacturing's performance and reliability. This means that Fraunhofer IFF makes workplaces smarter so that they assist workers according to their skills and maintain the quality of products and processes. They organize manufacturing and supply chains to be more energy and resource efficient. They consolidate regional energy, information and communications networks to make supply smart and reliable.

www.iff.fraunhofer.de

 # ALLIANCE-PROJECT

ALLIANCE aims at developing advanced research and higher education institution in the field of transport in Latvia by linking the Transport and Telecommunication Institute (TTI) with two internationally recognized research entities – University of Thessaly (UTH) and Fraunhofer Institute for Factory Operation and Automation (IFF).

The overall methodology is built around the analysis of the needs of Latvia and the surrounding region of the Baltic sea (Lithuania, Estonia, Poland) on knowledge gain about intermodal transportation terminals and the development of the tools to attain this knowledge, providing at the same time excellence and innovation capability. The analysis to be conducted during the first stages of the project, steps on the overarching relations among policy makers (e.g. government, city authorities), industry (e.g. transport operators, service providers) and education/research. Structured around three main pillars, organizational/governance, operational/services and service quality/customer satisfaction, ALLIANCE will deliver a coherent educational/training program addressed to enhancing the knowledge of current and future researchers and professionals offering their services in Latvia and the wider region.

The field of interest is "smart solutions and intermodal terminals", and the vision is that the knowledge transfer through twinning activities will benefit to creating a doctoral programme in Transport Economics and Management at TTI. The educational/training program is structured in three thematic areas thus, governance and policy development, smart solutions and decision-making framework.

http://alliance-project.eu/

This project has received funding from the *European Union´s Horizon 2020 research and innovation programme* under grant agreement No 692426

 # University of Thessaly

The University of Thessaly (UTH) was founded in 1984, and its main mission is the promotion of scientific knowledge through research and the contribution to the cultural and economic development of the local community and wider society. It is known for its excellent research performance and outstanding scientific achievements, in accordance with the international standards.

The present academic structure of UTH consists of sixteen Departments and four Faculties. Today, UTH has 9.467 undergraduate students, 1.471 postgraduate students and 1.148 PhD students. It also has 560 members of teaching and research staff, 98 members of teaching staff with a temporary teaching contract, 308 members of administrative staff and 57 members of Special Technical Laboratory Staff.

The Traffic, Transportation and Logistics Laboratory (TTLog) belongs to the Department of Civil Engineering of UTH. TTLog aims at the enforcement of the educational and research activities and the encouraging of close cooperation with other laboratories and research institutes. TTLog focuses on several scientific fields of research and educational activities, such as:

- Transportation planning, design and evaluation,
- Transport safety,
- User behavior analysis,
- Development and implementation of advanced transport technologies,
- Traffic and transport management,
- Freight transport and Logistics,
- Public transport,
- Transport interchanges.

http://ttlog.civ.uth.gr

TRANSPORT AND TELE-COMMUNICATION INSTITUTE, RIGA

The Transport and Telecommunication Institute (TTI) is the largest university-type accredited non-state technical higher educational and scientific establishment in Latvia. It was established in 1999 and is situated in Riga. Currently about 3000 students are enrolled in B.Sc., M.Sc. and Ph.D. programmes as well as in higher professional study programmes.

Main directions of academic activities are Electronics and Telecommunications, Information Technologies and Computer Science, Economics, Management and Business Administration as well as Transport and Logistics and Aviation Transport.

The Transport and Telecommunication Institute has its main research activities in:

1. ICT (Telematics)
2. Smart Solutions in Transport and Logistics
3. Digital Society and Economy

The TTI Research and Development Department has also:

1. Laboratory of Applied Software Systems (LAS) - The research and analysis are fulfilled using nowadays simulation software. The software of the laboratory allows doing the high-quality, representative and many-sided analysis of the research systems. Such projects as the projects connected with the new bus station in Riga, three level trestle of South bridge model and Liepaja city traffic macroscopic model can be mentioned as a vivid example.
2. Telecommunications, electronics and robotics center (TERC) - The center was founded in 2013 and includes nine laboratories equipped with the latest software and hardware widely used in academic and research activities. Each laboratory is a collection of contemporary technical, software and methodological maintenance, which allows conducting classes with students at the highest level.
3. TERC: Laboratory of Physics and Electrical Machines - The laboratory is equipped with training equipment of the company PHYWE, which allows students to explore the effect of the fundamental laws of physics. At the same time, the electrical machine equipment from the manufacturer K&H MFG, helps to understand the principles and work of modern electric motors.
4. Cisco Networking Academy - Cisco Company is the world leader in the area of network technologies. Cisco Networking Academy Program is aimed at the fundamental training of processionals for planning and operating computer networks using international standards. Training takes place in a special laboratory equipped with Cisco modern professional equipment.

The TTI has also a Telematics and Logistics Institute (TLI), which has been developed with the aim of making productive contributions to the continued progress of the transportation industry in Latvia by conducting applied research and development work in contemporary and future transportation issues.

www.tsi.lv

UNIVERSITY OF MISKOLC

The history of the University of Miskolc refers to Mining and Metallurgy back in 1735. Since those times, the organization of the University changed and was extended several times with new faculties, now being named since 1990 the University of Miskolc. While technical education has the longest tradition at the University of Miskolc, during the recent decades the institution was transformed into a true university. Currently it has eight distinct faculties. At present, faculties have more than 14000 students, who are assisted in their academic advancement by an educational staff of more than 700 and a non-educational staff of more than 800 members.

On most faculties, B.Sc. and M.Sc. programs are both offered for the students. The University of Miskolc started Ph.D. programs on the basis of accredited doctoral programs on October 1, 1993. Currently six Faculties of the University offer doctoral programs and award Ph.D. degrees in seven disciplines: Earth Science, Materials Science and Technologies, Engineering Science, Information Science, Law, Economics and Management Science, Literary Studies.

The University of Miskolc is the largest higher education institution in Northern Hungary. With its highly qualified experts, instrument infrastructure and laboratories, it contributes to scientific research and technical development in Hungary.

The Department of Materials Handling and Logistics is part of the Faculty of Mechanical Engineering and Informatics. The Department has a wide range of educational activities at 3 Faculties of the University of Miskolc in the frame of full time and part time trainings. The focus of research activities of the department lies in the following fields:

- Design of materials handling machines,
- Controlling and planning methods for modular materials handling systems,
- Computer integrated logistics, information logistics,
- Production and service logistics,
- Warehouse logistics, stock management,
- Recycling logistics,
- Maintenance and Quality assurance logistics,
- Global logistics, supply and distribution systems,
- Logistics management.

www.uni-miskolc.hu

NATIONAL AEROSPACE UNIVERSITY "KHAI", KHARKIV

The National Aerospace University, Kharkiv, Ukraine (KhAI) was established in 1930. Its history is closely connected with the development of aircraft engineering and science. The University is famous for its creation of the first European high-speed airplane with a retractable landing gear and the creation of the design of the turbojet engine. Nowadays, with 11000 students and 2700 academic staff, KhAI is one of the leading institutions of higher education in Ukraine in the training of specialists for aerospace industry in Ukraine and abroad. The University has trained more than 73000 engineers during its lifetime, 80% of KhAI graduates are among the specialists engaged in aerospace industry of Ukraine.

Many enterprises and institutions in the Ukraine use the scientific developments of the University: in 2004/2005 the University responsible bodies even developed more than 80 bilateral contracts aimed at the design, pilot development, testing and introduction of the products and technologies in 16 areas of aircraft design, machine building and related issues.

In 1994 KhAI has signed cooperation agreement with OvGU, which set different joint topics for educational and research collaboration in the sphere of aircraft design, developing the details out of composite materials, technologies and equipment for speed processing of steel construction etc.

www.efc.khai.edu

UNIVERSITY "MARTA ABREU" OF LAS VILLAS

The University »Marta Abreu« of Las Villas (UCLV) was founded in 1948 in Santa Clara. Approximately 7500 students are enrolled at the university, which consists of 12 faculties. The green, spacious campus is located on the outskirts and makes up its own small student town that may be reached by car, bus or train. UCLV is the third-biggest university of Cuba. It has ranked on top places in all national evaluations of the quality of teaching and research. UCLV is part of several national and international research networks and maintains scientific cooperation with 130 institutions around the world. Intensive collaboration with the OvGU in Magdeburg focuses on the departments of manufacturing engineering and quality management, mechanics, construction, computer science, automotive technology, process and environmental technologies and especially logistics and material handling systems.

The Department of Mechanical Engineering was founded in December 1959. The Department's most important fields of research pertaining to logistics and material handling systems are:

- Technical logistics,
- Quality management, quality engineering, metrology, measurement uncertainty
- Manufacturing (manufacturing engineering and welding technology),
- Environmental technology.

Furthermore, research is conducted in the fields of biomechanics, mechatronics, development and construction. The training is focused on mechanical engineering.

On February 2nd, 2007, the Department of Industrial Engineering and Tourism was founded. It is divided into two sections (Industrial Engineering and Tourism) and into two institutes: the Center for Tourism Research (CETUR) and Center for Business Management (CEDE). The central fields of research pertaining to logistics and material handling systems at the Department of Industrial Engineering and Tourism are:

- Quality management, quality engineering,
- Mathematical statistics, operations research, design of experiments, statistical simulation,
- Reliability and safety,
- Logistical networks.

www.uclv.edu.cu

LANDSHUT UNIVERSITY OF APPLIED SCIENCES

Landshut University of Applied Sciences offers 34 Bachelor's and Master's degree programmes in its six faculties of Business Administration, Electrical and Industrial Engineering, Computer Science, Interdisciplinary Studies, Mechanical Engineering and Social Work. Since its foundation in the year 1978, this University stands for excellent quality in the field of teaching and research. With 5,300 students, approximately 117 lecturers and a central campus the university has a manageable size. The Landshut University of Applied Sciences with its friendly premises full of light is a university of short distances. Its manageable layout permits intensive and personal assistance to the students. The university's practice-oriented teaching is directly focused on the needs of the world of business and the general society. Intensive co-operations between enterprises and the university within the framework of knowledge and technology transfer provide mutual impulses.

The Landshut University of Applied Sciences stands for the following main research topics:

- Lightweight Construction – materials, constructions and simulations
- Energy Systems – efficient networks, accumulating systems and installations
- Micro Systems / Embedded Systems – intelligent miniature systems
- Logistics and Production Planning – solutions for medium-sized enterprises
- Automotive – efficient automotive engineering and mobility
- Social Change – new concepts in the field of social work and health in the field of companies, organizations and the overall society.

The Technology Centre for Production and Logistics Systems (PULS) has the intention to increase the competitiveness of (mainly) small and medium enterprises from the region by means of three basic pillars: studies and further education, networking platform as well as applied research. As a part of the Landshut University of Applied Sciences, the Technology Centre PULS is a neutral partner and offers an ideal platform for the initiation, planning and execution of development projects in terms of innovative products. The core competencies of the Technology Centre PULS are in the following fields:

- Planning of manufacturing facilities
- Lean production
- Industry 4.0
- Lean logistics
- Workplace design
- Resource efficiency

www.haw-landshut.de

16

PKE Deutschland GmbH

PKE was founded in 1979 and is active both in Austria and internationally. The company is 100% privately operated and entirely in Austrian hands.

Over the last few years, PKE has grown from a medium-sized company to a corporate group with more than 1,100 employees and an annual sales volume of €220 million. The corporate headquarters of PKE is in Vienna and, with an additional 6 branch offices throughout Austria, PKE is able to offer its entire portfolio anywhere in the country.

In Europe, PKE is primarily focused on German-speaking neighboring countries, which are served by more than ten German and five Swiss branches, and on the growth markets in Central and Eastern Europe. We are on the ground in Poland, Czechia and Slovakia with our own subsidiaries—and, after a series of demanding showpiece projects, we are among the leading manufacturer-neutral solution providers.

The portfolio comprises:

- Security technology,
- Communication technology,
- Media technology,
- Electrical engineering,
- Building systems,
- Traffic engineering,
- Parking systems and
- Facility management.

By participating in national and international research programs, PKE's Development department also actively works on forward-looking concepts and technologies in its core business of security management systems, video systems, entrance management, control centers and holding cell communication.

PKE Deutschland GmbH   Landsberger Straße 187, D-80687 München   www.pke-de.com

ACADEMIC RESEARCH AND WRITING

THE DISSERTATION: WAY AND AIM

Prof. Dr.-Ing. habil. Prof. E. h. Dr. h. c. mult. Michael Schenk
Institute of Logistics and Material Handling Systems
Otto von Guericke University Magdeburg, Germany
Fraunhofer Institute for Factory Operation and Automation IFF, Magdeburg, Germany

Dr.-Ing. Dr. h. c. (UCLV) Elke Glistau
Institute of Logistics and Material Handling Systems
Otto von Guericke University Magdeburg, Germany

Dr.-Ing. Sebastian Trojahn
Institute of Logistics and Material Handling Systems
Otto von Guericke University Magdeburg, Germany

1   The Beginning

For every PhD student - whether a scholarship holder, a researcher at a university or employee at a company - the same essential questions come up over and over again (see Fig. 1):
- the question about the topic; the research question,
- the question about the efficient design of the weary way of finding and securing the answers and
- the question for the quality of scientific writing, which documents the results and proves their validity at the end of the research activity.

In the PhD regulations of the Faculty of Mechanical Engineering of the OvGU Magdeburg it says: "The doctoral thesis serves as proof of the ability to achieve results by independent scientific work, which contributes to the development of the scientific field, its theories and methods. The dissertation is a written, scientific work. It represents a performance based on this independent scientific research work.
The dissertation as a whole may not be published before the completion of the procedure. "[1]
Research means the "mental activity with the aim of gaining new insights in a methodical, systematic and verifiable way." [2] At the end there has to be a carefully formulated dissertation of 100 to 120 pages plus appendix. The quality of a dissertation is evaluated in proven, scientific practice by at least two independent scientists (usually university teachers with a doctoral degree), who are appointed by the faculty and asked for a written report on the dissertation. The procedure also includes a doctoral colloquium consisting of a presentation by the candidate followed by a public, scientific discussion with the doctoral committee and other persons interested in the subject.
With an age of about 60 years logistics is still a young discipline. However, referring to scientific work there is also a need to use the proven knowledge and practices from centuries of theoretical and practical scientific work from other scientific disciplines.
Indispensable features of any research work are to secure all data scientifically, the provability of the results and the proof of their validity. This aspiration must also be fulfilled by a dissertation. Heading to solve the problem the certainty (evidence) has to be developed and proven, so that it also opens up to a third party. The ultimate question is: How do I know that this is true what I say or write? How do I gain the certainty that the theories and methods I have developed have actually contributed to the development of my scientific discipline? How can I impart this certainty to a third party?

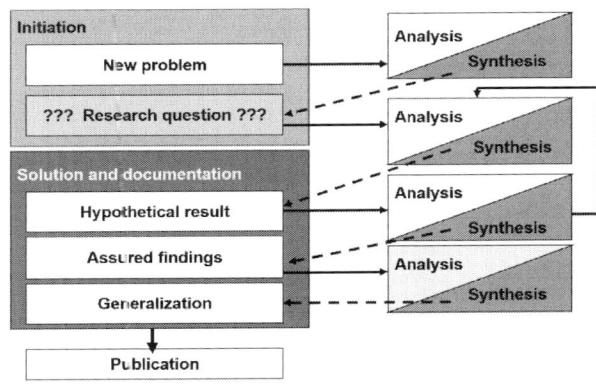

Figure 1: Procedure to solve a scientific problem

Figure 1 gives an overview about the main steps for dealing with research problems – also for PhD-thesis – which lead to scientific publications and dissertations. The left side of figure one shows the main phases with the most important outcomes. On the right you see the steps of development that always start with an analysis of the initial situation, the aims and requirements and lead to a search for potential solutions and methods to develop solutions. Thereby the tasks of analysis and synthesis change within the steps of development and require setbacks, critical, comparing evaluation and the scientific backup of the solutions.

2.1 Finding a Topic

In practice open questions of fundamental significance kick off scientific problem solving processes. The discovery and proof of gaps of knowledge in theory is the requirement to formulate a sustainable scientific question. The initiation can be made by the mentor when discovering gaps of knowledge, when finding conflicting results while writing a text book, when writing a report, when processing a project or out of a spontaneous idea (cf. [5]). The doctoral candidate can also find such gap of knowledge and bring this to a discussion with his/her advisor. For a proof of this "gap" an extensive analysis of the state of knowledge has to be done. The first phase of scientific writing is: examining, reading, structuring, discussing and framing.

Sources of information for doctoral candidates in logistics in Germany are e.g. logistic journals, textbooks, dissertations, publications of research institutions such as BVL, the VDI, relevant conferences, electronic databases and internet publications. The German journals are listed in [6]. Recommended is also the international literature as long as language barriers do not inhibit this. Thereby you may ask: How long has the literature to be backdated? As a rough guide in the field of logistics: 5 years for journals and dissertations, for reports of current scientific topics approx. 3 years, 10 years for textbooks and 3 years for conferences.

To start with the knowledge of experts publications is very precious. Your mentor can help here. It can also be helpful to assess the original source, for example: HAMMER for „Business Reengineering", IMAI for „Kaizen" or GOLDRATT for „Shortage Control". You should also clarify who is also researching on your topic and with what issue. The literature research from the start is convenient to extend the candidates expert knowledge. But the question for the "gap" should not be forgotten.

Formulating a research question is influenced by the "culture of asking questions" [14]. Generally spoken: "The scientific progress depends on a distinct formulation of unsolved problems." And "asking the right questions is an admirable skill." [7] A possible research question shall be illustrated on the example of organic mass logistics. It could be: "How should the ideal logistic system and the logistic processes of organic mass logistic look like for the supply of decentralized fuel cells?" The first answer to such a question is often: "I don't know!" It is important to estimate the resilience of such a research question. For this purpose it can help to answer questions like:

- Whom does this research question help?
- What scientific benefit is created by answering this question?
- What economic benefit is created by answering this question?
- What consequences can happen by failing to answer the question?
- What effort (material and time) has to be arranged?

For the topic of your dissertation it is important that "your fellow candidates topic is clearly defined" and to "frame your topic open for science" (cf. [3]).

Some examples for topics from the field of logistics are:

- Development of a method for a logistic risk analysis in production and supplier plants,
- Development of a procedure model for the model based strategy of the logistic network of a distributed production,
- Procedure for the decision on structure and location for decentralized plants.

The points of interest in logistics are logistic products (Services, Objects) logistic processes and logistic systems under the aspect of totality, effectiveness, performance, flexibility etc. as well as the development and evaluation of appropriate methods like e.g. analysis, scheduling, modelling, control and evaluation. Scientific research in logistics can refer to

- The search for a solution of a problem with no solution so far,
- An improvement towards the actual state,
- A systematization of knowledge of a greater field,
- A development of a new method,
- An usage of a method; also a synergetic conjunction of methods with new information in a field where these methods were not used before,
- A convergence towards an ideal state.

2.2 Possible Fields for Topics

The aim of a technical science is to gain knowledge and to embed this in the discipline. Typically there are e.g. tasks to create order through:

- Systematization,
- Range in historical order,
- Finding and defining categories.

Tasks for the development of methods like e.g.:

- Procedures (e.g. problem solving procedures),
- Models (explanation models, calculation models, working model).

Usage of models like e.g.:

- Gaining new insights,
- Creating of reference solution.

The difficulty of a research problem lies generally not only in its complexity but also that the solution and the suitable method have to be found.

Not only the kind of problem determines the procedure to solve the problem but also the method of evaluation of the results. For example has an improvement of the actual state to be evaluated in relation to the status quo, but a targeted solution can be evaluated best with a benchmarking process (for benchmarking analysis cf. [49, p41 to 51).

With regard to the quality of the process and the result of problem solving, these questions should be clarified at the beginning:
- When is my solution approach appropriate?
- When is my result appropriate?

This provides assurance (security) and enables goal-oriented and self-critical work. General quality requirements for research results are:
- New and innovativ,
- True (real) (in terms of reality) and verified (proven),
- consistent
- plausible und provable,
- generally valid,
- complete and comprehensive,
- integrable,
- extendable,
- feasible

3   To Plan the Argument

The second main step starts with the research question and serves to find solutions and solution ensurance (cf. figure 1). At the beginning, it is recommended to formulate working hypotheses. These hypotheses first reflect subjective, intuitive and therefore individual and very personal presumptions, assertions and justifications. Through validation, verification/falsification, certainty must be gained for these presumptions, assertions and justifications.

The proof of evidence (cf. figure. 2) must necessarily succeed (work out) and be provided. In technical sciences, validation and verification are often used as proof. These terms are derived from Latin. Validity describes the credibility of a model or an experiment and stands for the credibility of the theoretical result and its compliance with the real facts. Verification serves as a scientific validation of an assertion in the sense of proof or authentication (certification). There are different methods with which traditional proof is obtained. Regarding technology the actual state or an idealized target state, offer accepted standards of comparison. Comparisons based on measurements show how much better the new solution is than the already known state or how close it comes to an idealistic state. In the context of quality management the validation e.g. is used to verify if a product or service meets the requirements of the customer from a specific purpose of use.

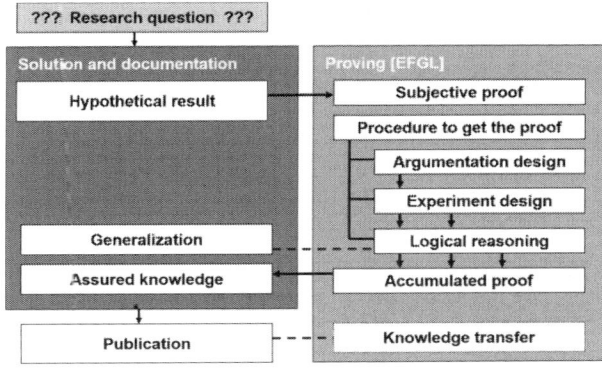

Figure 2: Create assured knowledge

3.1   Design of Argumentation

The design of argumentation includes the planning and execution of theoretical proof, e.g. for necessity of the research question, through logical conclusions or the discussion of advantages and disadvantages. Typical is also the use of solid terminology and logical lines of argument. It is always legitimate to cite authorities (known scientists or experts) and to lean on their arguments. The line of argument is traditionally made by abstract ideas, hypotheses, concepts and presumptions.

This abstract step is often made in conjunction with specific examples and by means of one's own logical thinking. Prerequisite for the initial, subjective check of proof is the experience of the doctoral student as a scientific expert.

3.2   Design of Experiment

The design of the experiment must be apt to "unambiguously accept or reject a hypothesis."[8] Therefore different types of experiences like empirical, sociological, technical and model-based have been developed. The design of an experiment usually includes multiple and staggered experiments, e.g. a statistical analysis of data supplemented by personally conducted interviews or an expert survey. Since the 16th century there are scientific rules for the statistical experimental design.

Names such as BACON and later of FISHER, TAGUCHI and SHAININ stand for procedures and proven methods that are recognized as "state of the art". (cf [4], p.63-74). Equally applicable are all standard valuation methods in logistics (cf [4], p.193-194), which question important aspects of a solution both qualitatively and quantitatively.

3.2.1   Enbedding

Finally, the embedding provides the control and proof of the fit(match) of the results and their fundamental validity in the researched academic discipline. It provides prove and provides indications to the degree of generalization of solutions and findings.

3.2.1   Accumulated Evidence

All three aspects (arguments, experiments and embedding) together generate the accumulated evidence (see Fig. 2), whereby there are connections between all aspects (see Fig. 3). These connections must be noted and must be consistent in order to form a harmonious whole.

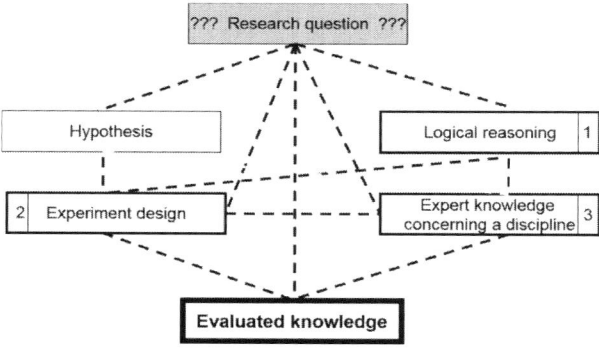

Figure 3: Three aspects of evaluated knowledge

After all, the findings of the scientific work are formulated suitably prepared and compacted (condensed) Outcome (result) is the "assured knowledge" (see. Fig. 2).

4   Creativity = Basic Requirement of Scientific Work

There is a variety of creativity techniques (see Tab. 1, 2) that promote systematic work. Particularly common are the well-known techniques:
- – Brainstorming,
- – Brainwriting and Brainpainting,
- – Synectic and
- – Morphologic Analysis.

Brainstorming and Brainwriting is assumed to be known and is not further explained.
Creativity is always related to intuition, interaction and leisure. Further, recommended literature on these creativity-related topics can be found at [17] to [20].

The synectics uses techniques of association which was already described by the Greek philosopher ARISTOTLE:
- – Similarities: e. g. Compliment / complement
- – Contradiction: e. g. vertical-horizontal
- – Spatial relationship: e.g. incoming goods
- – Quality check
- – Temporal relationship: e.g. before – after

The typical confrontations for the synectics are basically triggered through the following mind games and objects: pictures: e.g. Visual Synectic,
- – Side fields: e.g. side field integration,
- – Areas remote from the problem: e.g. bionics,
- – Emotive words: e.g. Extrem method,
- – Irritant Objects: e.g. catalog-technique,

- – Intuitive irritation: e.g. breaking up the search field

There are three different types of morphologies
- – *The conceptual morphology, in which the parameters are chosen in such a way that they sum up the concept in their entirety. It is used for constructions of all kinds and is especially well suited for product developments, in material handling technology and system planning in logistics.*
- – *The sequential morphology, for which a stepwise approach is characteristic. It is used for basic solutions, which are then successively refined.*
- – *The modified morphology (also "Attribute Listing"), in which for an existing basic concept variants are formed. These variants are usually formed of properties and characteristics*
- – *Starting from existing material or immaterial objects or processes, it is tried to systematically generate modifications*

With Morphological Boxes you can very well characterize an existing state, reveal options for an action and describe the new solution.
Table 1 shows well-known and recognized activities in the context of systematic creativity processes. Depending on the problem type, creative processes can also be differentiated as follows:
–Analytical problems
The problem structures should be recognized, possibly also the causes and the interaction. Exempels:
- – Which technical functions should a specific mode of transport have?
- – Which causes can lead to the failure of a function?
- – What measures can be taken in the event of a system failure?

The method of morphological analysis is best suited for this purpose:
– search problem
With the help of the application of search criteria, which are given by the definition of the problem, the quested (sought-for) wanted solution is selected from an existing or developed solution space.
The abundance of possibilities and their combination often makes solving the search problem difficult
Example:
- – Which single components can be used?

Suitable are the methods of Brainstorming as well as the methods of Brainwriting.
– Constellation problems
Existing knowledge is adjusted to new circumstances.
Example:
- – Conception of a new transfer device

Best suited for this is the method of synectics.

Abstract	To work out the substance from a problem, uncoupled from the specific
Alphabetise	To use the alphabet as a starting letter such as: A like…, B like…
Split	Split a problem in separate problems or modules to reduce complexity
Association	How would nature solve the problem? Link experience with examples!
Find relationships	Find a connection to the environment or to the elements of the system
Discuss, explain, formulate, inform	Talk to others, ask questions, find arguments, explain by using examples
Get inspired	What do the others do?
Integrate	Integrate in existing, known; build bridges
Combination	Join together (things, people…)
Criticise	Bring things into question
Start again	Start over with another view, forget all known
Open up	Extend the search, stop using filters
Break the poles	What's the third? Example: 1. Knowledge; 2. Ability; 3. Willing
Check	Verify, falsify
Creative thinking	What if…
Change roles	Change the point of view on a problem
Change view	Change perspective, role and think a problem through
Speculate	What would we never do under any circumstances?
Alter	Can a problem been altered?

Table 1: Task during systematic creative processing (1)

Generalise	Can a problem been solved in a more general way?
Simplify	Reduce complexity, can an easier problem been formulated and solved?
Sophisticate	Can a problem been blurred?
Alternate	Formulate the problem differently
Compare	Find differences and try to explain them
Mix	Mix up/ combine thought/solutions
Visualise	Pictures, sketches, diagrams
Look forward	How will it be done in future?
Look back	How was it done in the past?

Table 2: Task during systematic creative processing (2)

Problems of consequence
By following known legalities the problem will be solved.
Example:
-   Calculation of the throughput of a mining system.

Ideally suited are algorithms and logic chains.

-   Selection problem

Suitable alternatives will be examined with predefined criteria in respect of the aim of the task.
Example:
-   Selection of a control strategy

Most suitable are the techniques of "catalog-technique"

5 Publication = Inform and Delight Others

With a dissertation you not only record your research results and your authorship but also fulfil the evidence mediation upon any third party (cf. figure 1). Therefore it is necessary to show the proof of evidence (cf. figure 2) and all relevant coherences (cf. figure 3) namely in a logical and plausible way. In this connection it is helpful to use concrete examples to make something abstract more understandable [8]. The evaluation of any scientific publication or dissertation is divided in form and content.
The structure of the content should consider the following:
-   Motivation and Ambition,
-   Distinction,
-   Method,
-   Results and their degree of novelty,
-   Benefit for practical use,

- Benefit for science,
- Prospect on possible continues work.

For the linguistic design it is necessary to use a strict order and classification and also a clear, straightforward and descriptive language. Also there is a demand for explicitly defined terms, a consistent use following a strict and logical order (common thread) as well as a comprehensible argumentation.

You can find useful hints on effective writing of scientific publications here: [3], [9], [10]. LANG recommences in [9], p. 118 and 119, the following procedure:

First the abstract should be written, then the dissertation should be structured and topic sentences be formulated. Afterwards a sketch should be written in a single blow without correcting any mistakes. The revision should be an individual operation.

With only 120 pages in the main text part you have to eliminate minor matters. At a pinch these passages can be moved to the appendix.

It is helpful when formulating a text – also when having a writer's block – to ask yourself the following questions (see also [9], [10]):
- Who will you say something? (Addressee)
- What is my essence?
- How do your essences affect the reader?
- What kind of specialised knowledge can you expect from the reader?
- How does the knowledge have to be classified?

In the phase of writing the draft and in the phase of revision, it makes sense to get feedback from colleagues. Beside seminars for doctoral students, IFF-research colloquium and international workshops for doctoral students the daily breakfast and lunch suit well for this purpose (see [11], [12]).

For good information about university seminars see [13].

6 Honest Scientific Working – Against Plagiarism

Plagiarisms in dissertations are not just in public's eyes since the Gutenberg- scandal. The risk is always there when reading and analysing (so much) literature. Generally it is about "copyright violation" in the juristic sense. To develop the state of knowledge in a dissertation you have to read a lot of literature, critically question thoughts and opinions, classify what is already known…

Especially in theoretical literature dissertations it is important to work carefully whereby it has to be separated between mistakes (that should not happen but can happen) and intent (deceit). Forbidden is ghost-writing, fictional data, consciously "stolen" texts or pictures. In current literature it is expected that curtain software finds plagiarisms but on the other hand criticism is levelled against its effectiveness (see [14], [15]).

What is the difference between a proper literature dissertation and a plagiarism? This question was answered by a definition from 1930 (see [16]). To copy a chain of thoughts is called a "plagiarism of structure". When writing a scientific work and especially when writing a literature work, the author must always be aware that science has its own code of honour which is based on reliability; trust and honesty (see [14] and [16]).

7 References

[1] Otto von Guericke University Magdeburg. Promotionsordnung der Fakultät für Elektrotechnik und Informationstechnik und der Fakultät für Maschinenbau, zuletzt geändert durch die 2. Satzung zur Änderung der Promotionsordnung vom 20.03.2018

[2] Bundesverfassungsgericht Forschung & Lehre (2010). Hrsg. Deutscher Hochschulverband. Bonn 2. p. 81

[3] Bochmann, D. (2002): Vom Handwerk des Promovierens: Automatisierungstechnik 4. pp.187-191

[4] Illés, Béla; Glistau, Elke; Coello Machado, Norge I. (2007): Logistik und Qualitätsmanagement. 1. Auflage. Miskolc. ISBN 978-963-87738-1-4

[5] Voß. H.-G. (2004): Der Mensch, das neugierige Wesen. - Anmerkungen aus psychologischer Sicht. In: Forschung & Lehre 12. Hrsg. Deutscher Hochschulverband. Bonn. pp. 660-661

[6] http://www.fachzeitungen.de – Request from 13. February 2018

[7] Vollmer, G. (1995): Warum haben wir keine Fragekultur?-Wissenschaft lebt von Problemen. In: Forschung & Lehre 3. Hrsg. Deutscher Hochschulverband. Bonn. pp. 148-152

[8] Engelen, E.-M.; Fleischhack, C.; Galizia, C. G.; Landfester, K. (2010): Heureka -Oder: Wann jubeln Wissenschaftler? Evidenzgewinnung und -erzeugung im Forschungsalltag. In: Forschung & Lehre 11. Hrsg. Deutscher Hochschulverband. Bonn. pp. 810-812

[9] Lang, S. (2010): Strukturieren statt formulieren. Einfache Regeln, um eine wissenschaftliche Arbeit effektiver zu schreiben. In: Forschung & Lehre 2. Hrsg. Deutscher Hochschulverband. Bonn. pp. 118-119

[10] Kruse, O. (2008): Vertrackte Routine. Was tun, wenn das Schreiben stockt?. In: Forschung & Lehre 12. Hrsg. Deutscher Hochschulverband. Bonn. pp. 850-851

[11] Hinske, N. (1996): Eine Saat, die langsam wächst – Gesprächskultur und ihre Regeln. In: Forschung &Lehre 4. Hrsg. Deutscher Hochschulverband. Bonn. pp. 178-179

[12] Siefkes, D. (1995) Über die fruchtbare Vervielfältigung der Gedanken beim Reden.

Eine ökologische Theorie menschlicher Kommunikation. In: Forschung & Lehre 10. Hrsg. Deutscher Hochschulverband. Bonn. pp. 551-555

[13] Apel, H. J. (2001): Planlos und nach Gewohnheit? - Wie gestaltet man universitäre Seminare. In: Forschung & Lehre 3. Hrsg. Deutscher Hochschulverband. Bonn. pp. 138-140

[14] http://plagiat.htw-berlin.de - Request from 13. February 2018

[15] Weber-Wulff; D. (2011): Was leisten Plagiatserkennungssysteme. In: Forschung & Lehre. 2 Hrsg. Vom Deutschen Hochschulverband. Bonn. pp. 140-141

[16] Englisch, P. (1933): Meister des Plagiats oder Die Kunst der Abschriftstellerei. Hannibal-Verlag. Berlin-Karlshorst. 1933. pp. 81 f.Further literature according to creativity:

[17] Haupt, T. C. (2009): Intuition wird oft belächelt. Ist Kopf oder Bauch wichtiger für eine Entscheidung? In: Forschung & Lehre 7. Hrsg. Deutscher Hochschulverband. Bonn. pp. 520-521

[18] Funke, J. (2001): Kreatives Denken als Interaktionsprozess.- Zur Psychologie der Kreativität. In: Forschung & Lehre 5. Hrsg. Vom Deutschen Hochschulverband. Bonn. pp. 246-249

[19] Grigat, F. (1995): Denken mit der Stoppuhr?- Ein Plädoyer für mehr Mut zur Muße. In: Forschung & Lehre 6. Hrsg. Deutscher Hochschulverband. Bonn. pp. 319-321

[20] Radermacher, F. J. (1995): Kreativität - das immer wieder neue Wunder. In: Forschung & Lehre 10/1995. Hrsg. Deutscher Hochschulverband. Bonn. pp. 545-550

SCIENTIFIC PAPERS

A COMBINATION OF SIMULATION AND GENETIC ALGORITHM FOR SOLVING A STOCHASTIC INVENTORY OPTIMIZATION PROBLEM

Ilya Jackson, PhD student
Department of Mathematical Methods and Modelling
Transport and Telecommunication Institute, Latvia

PD Dr. rer. nat. habil. Juri Tolujew
Institute of Logistics and Material Handling Systems
Otto von Guericke University Magdeburg, Germany
Transport and Telecommunication Institute, Riga, Latvia

1   Introduction

Modern markets are extremely competitive. Businesses are facing unceasingly growing pressure on both prices and quality. Besides that, the company is required to swiftly respond to stochastic market conditions. Incorrect inventory policy leads not only to corporate losses, but also to overproduction. In this regard, traditional inventory policies are not appropriate anymore. Moreover, overproduction causes serious environmental problems, depleting natural resources and polluting the atmosphere.

The real-world inventory optimization is commonly characterized by the large-scale size and the necessity for nearly-optimal solutions in feasible computing times. That is why, the metaheuristics in general and genetic algorithms in particular are used so widely to define an optimal inventory policy. The world is full of uncertainty, which frequently makes classical deterministic approaches unsuitable due to excessive simplicity.

As it is mentioned in the recent research [1], real-life stochastic combinatorial optimization problems may be reformulated as a simulation in a natural way. Thus, the hybridization of metaheuristics and simulation techniques promises to be an efficient solution of stochastic inventory optimization and inventory control problems. First and foremost, the combination of simulation and metaheuristics is focused on efficiency taking into account stochastic components that may be contained either in the objective function or in the constraints. Such approaches are conventionally called simulation–based optimization or "simheuristics". The method aims to utilize a simulation instead of an objective function in traditional form and apply the genetic algorithm to find such simulation adjustments that would lead to the optimal output. In the proposed method, the iterative searching process of the genetic algorithm has to assess the quality of individual solutions, highlighting the promising ones. Besides, real-world stochasticity may be modelled throughout the best-fit probability distribution. The distribution may be either theoretical or empirical, without the need to be approximated to normal or exponential.

This paper describes a possible combination of discrete-event simulation and genetic algorithm to define the optimal inventory policy in stochastic multi-product inventory systems. The discrete-event model under consideration corresponds to the just-in-time inventory control system with a floating reorder point. The system operates under stochastic demand and replenishment lead time. The utilized genetic algorithm is distinguished by a non-binary chromosome encoding, uniform crossover and two mutation operators. The proposed approach contributes to the field of industrial engineering by providing a simple, but still efficient way to compute nearly-optimal inventory parameters with regard to risk and reliability policy. Besides, the method may be applied in automated ordering systems.

2   The simulation description

First of all, the method requires designing a simulation that corresponds to the real system with a high degree of accuracy. As it is already mentioned, such a simulation will play the role of an objective function. Thus, an optimization process will be reduced to the search of the best simulation adjustments. The inventory theory at its current stage has developed a significant mathematical foundation for solving problems related to the determination of the optimal inventory policy [2]. The most suitable model among considered is the model of Hopp and Spearman [3]. It is also worth noting that several distinguishing features were taken from "lost sales

(r, Q) inventory control model" [4]. The considered model makes several assumptions:
- Unfulfilled demands are defined as a lost opportunity and no backlog shall be fulfilled later
- Demand size, demand frequency and replenishment lead time are continuous random variables
- Product of a particular type is replenished by an individual supplier.

Discrete-event simulation paradigm is chosen in order to take into account random components without a dramatic increase in system complexity at the computational level. Unlike in continuous simulation, system dynamics is not unceasingly tracked during the simulation time. Discrete-event simulation contains a list of events, such that each event takes place at a particular instant of time altering the state of the system. It is important to emphasize that there are no changes in the system between consecutive events. That is why, the simulation laps in time from previous event to the next one and runs much faster saving precious computational resources. Each event is scheduled according to preliminary generated time t_n and executes sequentially. Generated time is appended to a time vector $T = (t_0, t_1 \ldots t_n)$, which may be interpreted as a time-counter. The total inventory assortment corresponds to the set of products P, such that each product $p_i \in P$. The storage capacity allocation is the first priority task. Presuming that I_{max} is the total storage capacity, we declare B as a vector of individual storage capacities assigned for products (Equation 1.).

$$\sum_{i=1}^{|P|} b_i = I_{max}; \ \forall \ b_i \in B \tag{1}$$

The simulation begins with an initial inventory level of I_p at t_0. During the simulation, emerging demands $x_{p,t}$ are are satisfied and the stock level declines gradually. If the stock level falls below a reorder point $r_{p,t}$, the inventory places a new order $y_{p,t}$ to refill the stock. Therefore, an inventory level at a particular moment of time equals to an inventory level in previous moment subtracting received demand and adding an order that was placed at $t - L$ Equation 2. Where Lp is the replenishment lead time for a product p. It is also worth noting that such a model aims to represent the inventory under some sort of just-in-time policy, thus, the order size $y_{p,t}$ equals to the corresponding maximal inventory capacity b_p subtracting the difference between the current inventory level $I_{p,t}$ and adjusted safety-stock SS_p Equation 3. In the proposed model, a new reorder point r_p is recalculated after each replenishment Equation 4. Where $m_{p,[t-L,t)}$ stands for a mean demand during the replenishment lead time and SS_p is a value of the corresponding safety-stock. Based on that, the number of arisen backorders for product p at time t may be determined as the step function Equation 5.

$$I_{p,t+1} = I_{p,t} - x_{p,t} + y_{p,t-L} \tag{2}$$

$$y_{p,t} = \begin{cases} b_p - (I_{p,t} - SS_p), if \ I_{p,t} > SS_p \\ b_p, if \ I_{p,t} \le SS_p \end{cases} \tag{3}$$

$$r_{p,t} = m_{p,[t-L,t)} + SS_p \tag{4}$$

$$O_{p,t} = \begin{cases} 0, \ if \ x_{p,t} \le I_{p,t} \\ x_{p,t} - I_{p,t}, \ if \ x_{p,t} > I_{p,t} \end{cases} \tag{5}$$

Discrete-event simulation of such models is simple enough and can be performed by the iterative algorithm (Figure 1).

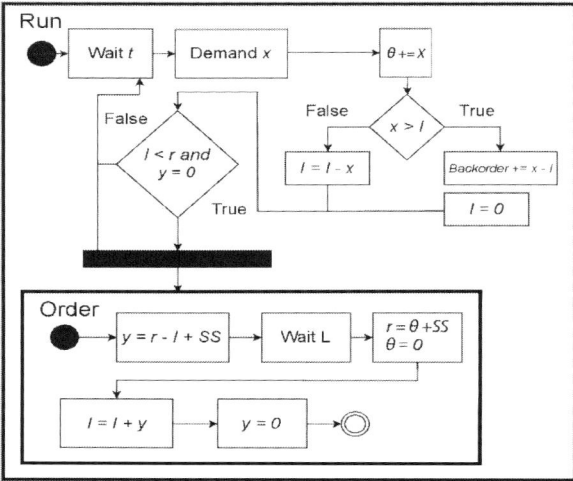

Figure 1: The logic behind the simulation

Each product in an assortment has a different market price and thus a different backorder cost o_p. Likewise, unit costs of storage and shipping, h_p and l_p respectively, vary depending on product's properties and subtleties of handling. Thereby, the total cost function for each product is the sum of the products of unit costs on number of units shipped, stored or backordered respectively Equation 6.

$$TC_p = l_p \sum_{i=0}^{t} y_{p,t} + h_p \sum_{i=0}^{t} I_{p,t} + o_p \sum_{i=0}^{t} O_{p,t} \tag{6}$$

According to Equation 3, the overflow may occur Equation 7.

$$F_t = \sum_{i=1}^{|P|} I_i > \sum_{i=1}^{|P|} b_i \tag{7}$$

Such a case may be taken into account by declaring a specific cost s related to the unit overflow and tracing the overflow level Equation 8. In real world, such a cost corresponds to the warehouse outsourcing or reverse logistics.

$$F_t = \begin{cases} 0, \ if \ \sum_{i=1}^{|P|} I_i \le \sum_{i=1}^{|P|} b_i \\ F_t, \ if \ \sum_{i=1}^{|P|} I_i > \sum_{i=1}^{|P|} b_i \end{cases} \tag{8}$$

In this regard the total costs function for an inventory as a whole TC is Equation 9.

$$TC = \sum_{p=1}^{|P|} TC_p + s \sum_{i=0}^{t} F_t \tag{9}$$

3 The optimization procedure

3.1 The optimization procedure

The genetic algorithm was invented and firstly introduced by Holland [5]. To date, genetic algorithms have been successfully implemented in logistics and supply chain management [6]. The motivation for combining genetic algorithm with simulation is that in real-life inventory problems, it is highly preferable to obtain a nearly-optimal solution for a precisely accurate model than the absolutely optimal solution for an oversimplified deterministic model. Genetic algorithms are totally different in comparison with the conventional search techniques. The optimization procedure starts with an initial set of randomly generated solutions that are called population. Each individual solution in the population is called a chromosome. The chromosomes undergo changes through sequential iterations. Such iterations are called generations. The chromosomes within the generation are evaluated, according to a fitness function. The next generation is composed by a set of new chromosomes, called offspring. Offspring, in its turn, is mainly formed by the fittest chromosomes, partially altered by either crossover or mutation operators.

In order to apply genetic algorithm, the following initial parameters are required:

- Population size (N) – the number of chromosomes in each generation
- Crossover rate (P_c) – the probability of executing a crossover operator
- Mixing ratio (P_u) – the probability for each attribute to be exchanged
- Mutation rate (P_{m1}) – the probability of executing a mutation operator 1
- Mutation rate (P_{m2}) – the probability of executing a mutation operator 2
- Mutation step ($delta$) – the gene-multiplier used by the mutation operator 2
- Tournament size (t)

3.2 Chromosome representation and fitness function

Practically, genetic algorithm is quite efficient in cases of large search space with lack of knowledge on the structure of the fitness function. The stochastic inventory optimization problem undoubtedly belongs to this domain. Moreover, in cases of high stochasticity, it becomes difficult to apply some traditional optimization techniques.

Genetic algorithm is quite famous as a problem-independent approach, nevertheless, the chromosome representation is a critical issue. Applying genetic algorithm to the inventory optimization problem under consideration, we are looking for such adjustments to simulation

parameters: storage-resources allocation B and corresponding safety-stock levels SS that lead to the best fitness. The chromosome may be encoded as a $|P|$ size list of integers $v = (b_1, SS_1, b_2, SS_2, \dots b_{|P|}, SS_{|P|})$. In such a list each odd element stands for the inventory capacity allocated to each product p and each even element represents adjusted safety-stock level for the corresponding product p (Figure 2).

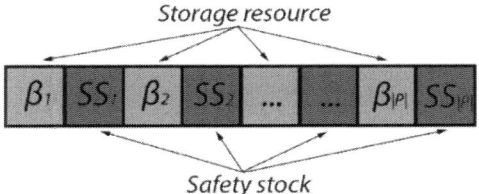

Figure 2: Chromosome representation

In such a simulation-driven approach, fitness function is evaluated by sequential runs of several simulations. In this case, fitness is the mean value of total costs calculated in several sequential simulation's runs. with the same parameters. We are looking for such parameters that lead to the minimal mean value of the total cost function Equation 8, satisfying the constraints Equation 9 and Equation 10.

$$\min_{a \in \mathbb{A}} E[\sum_{p=1}^{|P|} TC_p(a)] \qquad (10)$$

$$\sum_{i=1}^{|P|} b_i \leq I_{max}; \quad \forall i = 1, 2, 3, \dots, |P| \qquad (11)$$

$$SS_i \leq b_i; \quad \forall i = 1, 2, 3, \dots, |P| \qquad (12)$$

In case the solution does not satisfy constraints, the fitness will take extremely high values, due to infeasibility of such a solution. During the optimization procedure, such individuals (candidate solutions) will have only an insignificant chance to pass to the next generation.

It is pointing out that a suitable chromosome representation for the particular problem domain is an extremely important task, since a good choice will make the search faster and easier by restricting the search space. However, it is tremendously important to keep in mind that the crossover and mutation operators must take into account the design of the chromosome. It is important to emphasize that in the considered problem a non-binary chromosome representation was chosen.

The main reason why binary representation is the most frequent is the simplicity to implement and popularity in academic papers. Moreover, binary chromosome representation is usually space-efficient, that is why it was so popular in times, when memory was a serious problem. However, in real-world problems, it becomes common to create a genotype representation that corresponds to the considered problem with a high degree of accuracy.

3.3 Crossover and mutation

Crossover is the distinguishing operator of the genetic algorithm. Basically, it is a process of taking two parent solutions and production of offspring solutions in order to get a new, potentially better one. Crossover is used to vary chromosomes from one generation to the next. In order to solve the stochastic inventory problem, the uniform crossover is proposed (Figure 3).

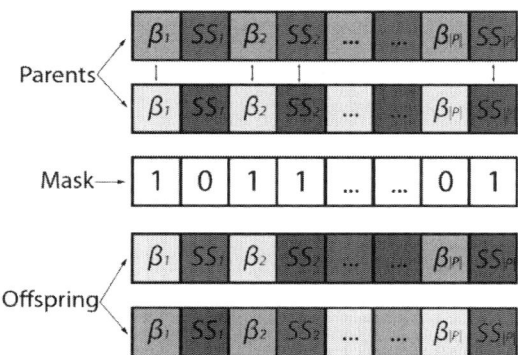

Figure 3: Uniform crossover representation

In the uniform crossover individual genes in the chromosome are compared between two parents and swapped with the fixed mixing ratio P_u. Uniform crossover is chosen for two main reasons. Firstly, since genes in the chromosome correspond to different simulation parameters SS and B, we seek a way to keep odd and even genes separated. Secondly, the uniform crossover is an efficient way to avoid the premature convergence [7].

```
Pu ← probability of swapping values
v⃗ ← first vector ⟨v₁, v₂, …, vₙ⟩
w⃗ ← second vector ⟨w₁, w₂, …, wₙ⟩
for i in (1, length of vector) do
    if Pu ≥ random number then
        swap the values of vi and wi
return v⃗ and w⃗
```

Besides, genetic algorithm requires a mutation operator to perform the optimization. Taking into account the particularities of chromosome encoding, it is proposed to apply two different mutation operators ("mild" and "radical"). The radical mutation is applied in order to prevent the premature convergence (otherwise population may get stuck in local optima). In radical mutation we replace gens in the chromosome by a new integer number in a feasible range (0, I_{max}) with the probability P_{m1}.

```
Pm1 ← probability of replacing
v⃗ ← vector
```

```
for i in (1, length of v⃗) do
    if Pm ≥ random number then
        vi ← random feasible integer
return v⃗
```

On the other hand, mild mutation is applied to accelerate convergence. The mild-mutation operator alters genes in the chromosome with the probability P_{m2} by multiplying them on some relatively small step *delta* rounding to the nearest integer after that.

```
Pm2 ← probability of altering value
v⃗ ← vector
for i in range (1, length of v⃗) do
    if Pm ≥ random number then
        vi ← round(vi * delta)
return v⃗
```

3.4 Selection

It is concluded by Miller and Goldberg [8] that tournament selection is an efficient and robust mechanism for working with imperfect fitness functions. Tournament selection runs several "tournaments" among *t* individuals (chromosomes) randomly chosen from the population. The fittest individual in each tournament is selected for the following crossover. Since weak individuals have relatively a small chance to be selected in large tournaments, it is quite important to find the optimal tournament size *t*. Tournament Selection can be programmed by the extremely simple algorithm:

```
P ← population
t ← tournament size, t ≥ 2
Best ← random individual from P
for i in range (2 to t) do
    Next ← random individual from P
    if Fit(Next) > Fit(Best) then
        Best ← Next
return Best
```

Tournament selection has several significant benefits over alternative selection methods, namely, it is both simple and efficient to code, it works with parallel architectures and, lastly, it may be easily adjusted.

4 Conclusion

In conclusion, the proposed optimization technique is a simple to design and computationally efficient approach to find nearly-optimal inventory policy in stochastic multi-product inventory systems. Additionally, the combination of discrete-event simulation and genetic algorithm provides a flexible method to solve complex problems with lack of knowledge on the structure of the objective function. Besides, the key advantage of such a simulation-driven approach is the possibility to trace inventory dynamics in details. It is supposed that the method may be applied in automated ordering systems by retail companies.

The research also concludes with a statement that the non-binary chromosome encoding in combination with uniform crossover and two mutation operators provide a fine balance between convergence speed and likelihood of premature convergence. There are still several minor problems to solve, such as the program-optimization of both the simulation and genetic algorithm. Moreover, it is crucially important to test the proposed approach on problems with higher dimension and compare it to alternative metaheuristic techniques. These issues are waiting to be deeply explored in a future research.

5 References

[1] Juan, A.A.; Faulin, J.; Grasman, S.E.; Rabe, M. and Figueria, G. (2015): A review of simheuristics: Extending metaheuristics to deal with stochastic combinatorial optimization problems. Operations Research Perspective 2:62–72.

[2] Zipkin, P.H. (2000): Foundations of inventory management. McGrawHill. 8. Hopp, W. H. and Spearman M. L. (2008) Factory Physics. Waveland Press.

[3] Hopp, W. H.; Spearman M. L. (2008): Factory Physics. Waveland Press.

[4] Kouki, C.; Jemai, K.; Minner, S. (2015): A Lost Sales (r,Q) Inventory Control Model for Perishables with Fixed Lifetime and Lead Time. Int. J. Production Economics 168:143-157.

[5] Holland, J.H. (1975): Adaptation in natural and artificial systems. Ann Arbor, MI: University of Michigan Press.

[6] Yeh, W.C.; Chuang, M. C. (2011): Using multi-objective genetic algorithm for partner selection in green supply chain problems. Expert Systems with applications 38(4):4244-4253. 15. 159-177.

[7] Michalewicz, Z. (1996): Evolution strategies and other methods. In: Genetic Algorithms + Data Structures = Evolution Programs. Heidelberg, 159-177. 1. Miller, B.L. and Goldberg, D.E. (1995) Genetic algorithms, tournament selection, and the effects of noise. Complex systems, 9(3), pp. 193–212. DOI:10.1162/evco.1996.4.2.113.

[8] Miller, B.L.; Goldberg, D.E. (1995) Genetic algorithms, tournament selection, and the effects of noise. Complex systems 9(3):193-212.

A SHORT SURVEY OF IMAGE PROCESSING IN LOGISTICS

Dipl.-Sporting Dipl.-Ing. Hagen Borstell
Institute of Logistics and Material Handling Systems
Otto von Guericke University Magdeburg, Germany

1 Introduction

Logistics is the science of comprehensive analysis, planning, design, control and monitoring of spatiotemporal transformation processes of goods, persons and related information [1]. Due to increasing global diversity of goods and globalization of markets, the service providing logistics systems are becoming more complex and challenging. This results in increasing needs for standardization, automation and digitalization of both the logistics processes and the associated flow of information. The concept of smart logistics zones was created in response to these demands [2]. A smart logistics zone is the domain of (smart) logistics objects in a smart logistics infrastructure in which logistics processes run as efficiently as possible. The smartness is implemented by means of sensor modules, which are attached to logistics objects or infrastructure components [3]. In this paper we focus on sensors that are sensitive to electromagnetic waves (e.g. infrared, visible light, thermal) and provide images. Images are 1D, 2D or 3D numerical representations of spatially distributed physical characteristics of objects or environments. Images can be produced either by scanning or by direct imaging. A sensor that captures an image sequentially by moving sensor parts or itself is called a scanner. If a sensor is capable of capturing an image without scanning, it is called a camera. The advantage of cameras compared to scanners is that image capturing can be carried out very quickly and it follows that image sequences can be recorded and analyzed. It should be mentioned that cameras can be used as a building block for scanners. e.g. depth cameras can scan a full 3D environment by performing multiple shots at different locations [4]. By means of image processing both the images from cameras and scanners can be further processed to extract relevant data.

In this article, applications and trends of image processing in logistics will be presented. The term image processing refers to the entirety of systems that capture images and operations that are applied to images and either produce enhanced images or extract data from the images. The information captured by image processing is used to control or monitor logistics processes and thus contributes to the smartness of smart logistics zones.

2 Applications of image processing in logistics

A literature survey was conducted to identify image processing applications in logistics. Subsequently, a categorization was performed. As a result, various categories of logistics applications of image processing emerged:

- Traceability and trackability
- Volumetric properties of goods
- Inspection and quality control of goods
- Equipment condition monitoring
- Occupancy of storage and traffic areas
- Security and protection of infrastructure
- Process modelling and simulation
- Manual picking and packing
- Manually-guided handling systems
- Automated handling systems
- Visual documentation and monitoring

In some applications, image processing directly influences the handling and flow of goods. In other applications, image processing is involved in collecting information for decision-making along the supply chain. Also, there are applications that provide a visual data flow, which is subsequently interpreted by humans. From this point of view, image processing supports both the automation of logistical processes and human operators in logistics systems with cognitive, physical and visual assistance functions. In the following sections, the above categories are outlined in more detail. The paper closes by presenting some important trends of image processing in logistics.

2.1 Traceability and trackability

Traceability and trackability concerns the identification and localization of logistics objects such as goods, containers, vehicles or persons within logistics systems. It is implemented by attaching optical codes to the objects (e.g. barcode, data matrix code, QR-code, light pattern or characters) and by image processing units capturing and reading the optical codes. Identification systems are widely used to identify incoming and outgoing goods in warehouses and distribution centers. In recent years, an increasing number of camera-based systems have been put to practice, since they are capable of reading 2d codes or optical character code (OCR) in addition to 1D barcodes [5]. This applies to both manual and fully automatic identification systems [6]. Besides warehouses and distribution centers, OCR systems are widely used in port logistics to read container codes and license plates of trucks [7]. Furthermore, the camera approach enables new products, such as the ProGlove for hands-free documentation [8].

In addition to identification, camera-based systems allow accurate localization of codes relative to a camera. Based on this principle, there are two different approaches to localization systems: Either cameras are mounted on the vehicle, e.g. forklifts, to localize codes in the environment [9] or cameras are placed in the work environment to localize codes on forklifts [10]. The first variant is called self-localization and advantageous if privacy protection is required. The second variant is called target-localization and advantageous if cameras are already installed and can be used for localization. With camera-based systems, not only forklifts but also packages or pallets could be localized, provided they can be equipped with optical codes [11]. Laser scanners can be used for self-localization as long as the objects are large enough to hold a scanner. If that's the case, forklifts can be localized by analyzing the distance measurements relative to the walls [12].

New approaches to identify and localize logistics objects avoid using optical codes, but rather use either convolutional neural networks or data fusion methods. Given objects are optically clearly distinguishable, deep learning methods can be successfully applied [13]. If objects such as pallets are unambiguously related to handling units that are localized, e.g. a forklift, data fusion can be a promising approach [14].

2.2 Volumetric properties of goods

Volumetric properties of goods are important planning and billing quantities in logistics processes. Therefore, scanner-based dimensioning systems are widely used to detect parcel dimensions on conveyor systems [6]. Since palletized goods are usually handled manually, the degree of automation in dimensioning palletized goods is also low [15]. Increasingly, attempts are being made to reduce measurement time of dimensioning palletized goods by enabling transport-integrated measurement [16]. Depth cameras based on time-of-flight or structured light are well suited for those measurements, because they are capable of capturing a complete scene with a single shot [17]. Currently, these approaches are still limited to use cases that are outside the scope of legal metrology (e.g. EU Directive 2014/32/EU). However, due to cost advantages of camera-based systems over scanner-based systems [3], other fields of application attract notice, such as mobile parcel dimensioning or product data acquisition [18].

Standard cameras are also suitable for dimensioning of packages or palletized goods, provided the cameras are calibrated and reference points are specified by an operator [19]. In the field of bulk materials measurement, scanner-based laser systems are often advantageous because they are robust to external light sources [20]. However, attempts are being made to evaluate and establish camera-based systems for bulk measurement as well [21].

2.3 Inspection and quality control of goods

Ensuring that the quality of goods is maintained is an important issue along the supply chain. Of great importance in quality inspection are optical systems that are integrated into manufacturing processes of goods. However, these systems are not in the scope of this paper. But also in the logistics part of the supply chain, optical systems are used to ensure the quality of the goods. Scanner-based dimensioning systems are used in automatic conveyor systems to detect defects on incoming parcels [6] or containers [22]. Also, camera-based systems are suitable to detect anomalies on parcels [23]. Furthermore, documentation and monitoring system are used to prove integrity of goods (section 2.11).

2.4 Equipment condition monitoring

Image processing can contribute to condition monitoring of logistics equipment. By using thermography cameras, which provide temperature images, defects on conveyer systems resulting from high friction can be detected [24]. Furthermore, depth cameras were used for condition monitoring on belt conveyors in order to detect belt displacement at an early stage [25]. Obviously, image processing systems offer advantages, especially for condition monitoring of long belt conveyors, which otherwise would have to be checked manually.

2.5 Occupancy of storage and traffic areas

Objects occupy space: pallets are stored in warehouses, vehicles are parked in parking lots and packages are loaded onto delivery trucks. The availability of logistics areas is important for efficient storage and transportation processes. Therefore, camera-based approaches have been developed to determine free capacity of transport vehicles. One approach is the use of cameras, which are oriented towards trailer gates [26]. Here, manually specifying reference points in images of the calibrated cameras are sufficient to determine free loading meters. Another approach is to use depth cameras to determine free loading meters. Either the depth cameras can be mounted in the vehicles [27] or outside the vehicles at the gates and oriented towards the trailers [28]. Knowing current loading conditions, space and operating costs can be optimized.

The detection of free parking lots for vehicles is an important component of smart city solutions. Cameras and convolutional neural networks can be used to detect if a certain parking lot is occupied by a car [29]. This approach can also be transferred to other domains, e.g. detection of free storage bins in warehouses or transshipment ports. The occupancies of storage bins can be used to detect critical process situations or to optimize transport routes [30]. By detecting occupancies of docks or parking lots, this technology can support dock and yard management. Furthermore, loading states of forklifts are of interest, e.g. in order to determine the coordinates of delivered goods in warehouses [31]. Neural networks, as used in parking lot monitoring, could lead to acceptable results.

2.6 Security and protection of infrastructure

Without infrastructure there is no logistics: transport routes as well as logistics facilities form the basis for logistics processes. Hence, they must be protected against external or internal hazards, e.g. accidents, theft or terror attacks. Video surveillance is an effective solution to increase security in logistics facilities. Available solutions are constantly being extended by new functions, e.g. intrusion detection, people counting or face recognition. Due to legal requirements, privacy features are becoming increasingly important [32]. The increasing amount of data associated with security and protection functions render visualization extremely important in security applications [33]. In the domain of passenger handling, for example, camera-based systems for people counting [34], crowd density estimation [35] and crowd behavior analysis [36] are used to increase security. With the help of forecasting functions, critical situations can be predicted, and preventive measures initiated. In the domain of freight transport, camera-based systems are capable of computing a freight fingerprint for the detection of potential security breaches along the transport chain [37].

2.7 Process modelling and simulation

Simulation methods are used to plan and predict performance of logistics systems. Image processing can be helpful in collecting process data for close-to-reality simulation. By means of image processing methods (e.g. object recognition or object tracking) material flow parameters such as number, state, flow directions, throughput or throughput time can be calculated. Subsequently, these parameters can be used for simulation-based evaluation and optimization of logistics processes [38][39].

2.8 Manual picking and packing

Although there is a significant trend towards automation in logistics, the proportion of manual work in handling processes is still very high. This is due to the diversity of goods to be handled, which makes automation very difficult. Since manual work is very error prone, especially in picking and packing operations, assistance systems are being developed, with which error rates can be reduced.

Within picking assistance systems, depth cameras are used to detect steps of moving of the picking operations [40] or color cameras are used to detect objects directly [41]. Both variants can be used to determine whether an object has been correctly placed into or has been removed from a container, and thus whether an error has occurred or not. Providing workers with visual feedback about the handling process is a way to avoid the occurrence of errors. Process information can be projected directly into the working environment using visual projectors [42]. Also, within packing assistance systems, depth cameras for motion analysis or color cameras for object recognition are being used. In the first case, the movement of hands is a criterion of whether the correct goods were packed [43]. In the second case, goods are recognized directly in the package, they are compared to order lists, and it is checked whether particular goods belong to a package [44]. New developments in the field of deep neural networks are currently opening up new application in the context of picking, packing and placing. In retail, these technologies are being tested to establish stores without checkout lines [45].

2.9 Manually-guided handling systems

Due to their weights, goods are often moved with powered handling systems. Navigation or maneuvering is still performed by humans, supported by assistance systems. A common assistance system for transportation with forklifts, trucks or cranes are rear view or surround view systems [46]. Cameras are also mounted on vehicles with moving machine parts (e.g. forklift masts) to assist operators with close-up views, e.g. pallet handling in high-bay storages. The images of these cameras can also be analyzed with respect to distance or position, and handling instructions can be derived [47]. Furthermore, obstacles can be detected and warnings can be generated [48].

Automated-guided vehicles (AGVs) are discussed in the next section. However, AGVs have also been extended by human-guided handling capabilities. More precisely, AGVs can be controlled by human gestures, which can be captured by depth cameras. Based on detection of human-gestures, assistance functions have been implemented to trigger and accompany transport [49] and handling [50] processes.

2.10 Automated handling systems

Automation of material handling systems includes, for instance, handling, loading, picking and sorting operations along the supply chain. Image processing can support this automation. For unloading operations with robots, the individual packages, sacks or other object types must be recognized and localized. This can be achieved by using depth cameras and methods of object recognition and pose estimation [51]. Similar applications are de-palletizing of goods [52] and robot-based order picking in warehouses or distribution centers [53]. Such technologies are also suitable to control conveyors, which perform complex intralogistics tasks such as package layering or singulation of packages [54][55]. Laser scanning systems for collision protection and camera-based systems for optical track guidance have long been used successfully in automated guided vehicles. Recent research is devoted to obstacle detection for safer and more flexible man-machine environments [56] and for optimization of vehicle routes [57]. In addition to laser-based systems, camera-based systems are increasingly being used in these applications, in particular depth cameras are combined with modern object recognition methods [58].

2.11 Visual documentation and monitoring

Logistics is a network of machine operations and human actions. Humans apply action to the process in response to current conditions. One way of communicating process states to humans is through visualization. Visualization based on image data can contain complex content, which is, however, very easily decipherable by humans. Visual documentation of liability transitions can reduce liability risks and improve risk management [59]. When operating large areas such as logistics hubs, passenger terminals or event areas, visual monitoring systems based on heterogeneous sensor systems are ideal for the operator to assess the situation at a glance [60].

3 Trends of image processing in logistics

Trends and their impact are difficult to predict. However, the current literature review reveals that there are certain technologies and methodologies that are being used to an increasing extent:

- **Depth Perception**: The availability of consumer depth cameras triggered the development of a variety of new systems, e.g. systems for mobile package dimensioning, container unloading, picking assistance or flexible conveyor controlling.
- **Embedded Vision**: Image processing units are increasingly designed as Internet of Things (IoT) nodes. They are equipped with communication modules and do not deliver pictures, but state data of processes.
- **Deep Learning**: Deep learning methods are increasingly used for object recognition in logistics, e.g. robot picking or grab-and-go

groceries. In particular, an attempt is made to supersede code-based ID readers.
- **Cameras**: Improvements of quality features (e.g. resolution, privacy protection) lead to greater integration of cameras into logistics processes. In addition to video surveillance systems for visualizing processes, cameras are also increasingly being used for data collection instead of scanners, e.g. identification, dimensioning.

Beyond that, tendencies such as the use of software frameworks and the application of methods of digital engineering (e.g. virtual operation) can be identified. The future will show which of these tendencies and trends will make the largest contribution to digitalization of logistics and to smart logistics zones respectively.

4 Summary

In this publication, applications and trends of image processing in logistics have been presented. The identified categories of logistics applications of image processing illustrate the significant contribution of these technologies to digitalization of logistics and to the concept of smart logistics zones. Some innovations within categories, such as (a) volumetric properties of goods, (b) manual picking and packing and (c) automated handling systems are leading the way within new technological trends. Which applications and which trends will gain in importance must be further monitored.

To determine base functions of image processing and their special characteristics in terms of logistic applications represents an important aspect of future research. Base functions, such as pose detection, should act as the smallest building blocks for image-based sensor systems, supporting, in particular, the planning and implementation of those systems for logistics applications.

5 References

[1] Krampe, H. et al. (Hrsg.) (2012): Grundlagen der Logistik: Theorie und Praxis logistischer Systeme. 4. Auflage. München: HUSS-VERLAG GmbH.

[2] Schenk, M. et al. (Hrsg.) (2015): Produktion und Logistik mit Zukunft : Digital Engineering and Operation. Heidelberg: Springer Vieweg.

[3] Wegner M. et al. (Hrsg.) (2013): Low-cost Sensor Technology - a DHL Perspective on Implications and Use Cases for the Logistics Industry. Troisdorf: DHL Customer Solutions & Innovation.

[4] Kersten, T. et al. (2016): Genauigkeitsuntersuchungen handgeführter Scannersysteme. In: Dreiländertagung der DGPF, der OVG und der SGPF in Bern, Schweiz - Publikationen der DGPF, Band 25, pp. 271-287.

[5] GS1 Germany (2015): 2D symbols in distribution and logistics. url: https://www.gs1.org

[6] Vitronic Dr.-Ing. Stein Bildverarbeitungs-systeme GmbH (2012): Technische Dokumentation - Kamerabasierte Daten-erfassung für Warehouse & Distribution. url: https://www.vitronic.de

[7] Shetty, R. et al. (2012): Optical Container Code Recognition and its Impact on the Maritime Supply Chain. In: Proceedings of the 2012 Industrial and Systems Engineering Research Conference. Florida, USA.

[8] Kirchner, T. (2016): ProGlove launches their first smart Glove for Industry. Workaround GmbH (ProGlove). url: http://www.proglove.de

[9] Jung, M. et al. (2015): An Accurate and Efficient Camera-based Indoor Positioning Approach for Intralogistic Environments. Proceedings of the XXI International Conference MHCL'15. Belgrade: University of Belgrade, Faculty of Mechanical Engineering, pp 133-138.

[10] Borstell, H. et al. (2013): Vehicle Positioning System based on Passive Planar Image Markers. In: Indoor Positioning and Indoor Navigation (IPIN), 2013 International Conference on, pp. 1–9.

[11] Lewin, M. et al. (2017): Optimization of Production-oriented Logistics Processes Through Camera-based Identification and Localization for Cyber-physical Systems. In: Advances in Production Management Systems, Springer, pp. 168-176.

[12] IdentPro GmbH (2018): Staplerleitsystem identplus. url: http://identplus.net/de/das-3d-staplerleitsystem-identplus/

[13] Schlüter, M. et al. (2018): Vision-based Identification Service for Remanufacturing Sorting. Procedia Manufacturing, 15th Global Conference on Sustainable Manufacturing 21, pp. 384–391.

[14] Borstell, H. et al. (2014): Pallet Monitoring System Based on a Heterogeneous Sensor Network for Transparent Warehouse Processes. In: Sensor Data Fusion: Trends, Solutions, Applications (SDF), pp. 1-6.

[15] AKL-tec (2017): Datenblatt APACHE portal first choice - Der preiswerte Einstieg in die APACHE Frachtvermessung. url: https://akl-tec.de/produkt/apache-portal-first-choice/

[16] AKL-tec (2017): Datenblatt APACHE flying forklift - Eine neue Dimension für das Messen und Verwiegen „on the fly". url: https://akl-tec.de/produkt/apache-flying-forklift/

[17] Borstell, H. et al. (2013): Prozessintegrierte Volumenerfassung von logistischen Palettenstrukturen auf Basis von Low-Cost-Tiefenbildsensoren. In: 3D NordOst 2013 - Tagungsband, Berlin, pp. 115–124.

[18] Fechteler, M. et al. (2016): Prototype for Enhanced Product Data Acquisition based on Inherent Features in Logistics. In: 2016 IEEE 21st International Conference on Emerging Technologies and Factory Automation (ETFA). pp. 1–4.

[19] VLS Engineering GmbH (n.d.): Messen im Bild. url: https://v-l-s.com/messen-im-bild/

[20] LoadScan (n.d.): Loadscan OverView - Load Management Solutions. url: https://www.loadscan.com/solutions/overview

[21] Richter, K. et al. (2012): Robust Depth Imaging Measurement for Bulk Materials Technology. Cement International 01/2012, pp. 66–75.

[22] Bundesministerium für Wirtschaft und Energie (2016): Transportschäden sicher identifiziert. url: https://www.zim-bmwi.de/erfolgsbeispiele/transportschaeden-sicher-identifiziert/

[23] Noceti, N. et al. (2018): A Multi-camera System for Damage and Tampering Detection in a Postal Security Framework. J Image Video Proc. 2018, pp. 11.

[24] Wenzel, S. et al. (2011): Condition Monitoring in Logistics – a New Approach for Maintenance. In: ICEME 2011- 2nd International Conference on Engineering and Meta-Engineering, 27-30 March 2011, Orlando, USA.

[25] Cao, L. et al. (2013): Kompakter, multifunktionaler Sensor auf Basis von Tiefenbildtechnologie für die Schüttgut-technik. In: 13./14. Forschungskolloquium am Fraunhofer IFF Forschung vernetzen - Innovationen beschleunigen, pp. 105–112.

[26] VLS Engineering GmbH (n.d.): Messen im Bild jetzt auch für Lademeterermittlung. url: https://v-l-s.com/messen-im-bild-lademeterermittlung/

[27] Borstell, H. et al. (2015): Toward Mobile Monitoring of Cargo Compartment Using 3D Sensors for Real-Time Routing. In: Logistics Management, Lecture Notes in Logistics. Springer, Cham, pp. 189–200.

[28] ZIH Corp (2017): Solution Guide: Making Your Loading Operations Smarter and More Connected. url: https://www.zebra.com

[29] Amato, G. et al. (2017): Deep learning for decentralized parking lot occupancy detection. Expert Systems with Applications 72, pp. 327–334.

[30] Borstell, H. et al. (2013): Echtzeitinformatio-nen in sicherheitskritischen GIS-Anwendun-gen. In: GeoForum MV 2013 - Neue Horizonte für Geodateninfrastrukturen - Open GeoData, Mobility, 3D-Stadt. pp. 181–190.

[31] Özgür, Ç. Et al. (2016): Comparing Sensor-based and Camera-based Approaches to Recognizing the Occupancy Status of the Load Handling Device of Forklift Trucks.

[32] KiwiSecurity (2016): KiwiVision - Privacy Protector. url: https://www.kiwisecurity.com

[33] Borstell, H. et al. (2013): Virtual top view: Towards real-time aggregation of videos to monitor large areas. In: Computer Analysis of

Images and Patterns. 15th International Conference, CAIP 2013. Vol.2. pp. 546–554.

[34] Kempe, M. et al. (2011): Echtzeitnahe Analyse des Personenaufkommens in öffentlichen Bereichen. 9./10. IFF-Kolloquium Forschung vernetzen - Innovationen beschleunigen Magdeburg, pp. 85–90.

[35] Borstell, H. et al. (2012): Image-Based Situation Assessment in Public Space. In: Future Security, Communications in Computer and Information Science, Berlin: Springer, pp. 61–64.

[36] Burrows, P. (2015): Measuring the customer's journey through London City Airport. In: Journal of Airport Management 9, pp. 103–108.

[37] Borstell, H. et al. (2012): Security in Supply Chains in the Scope of Surface Transport of Goods by Secure Information Patterns on the Freight - Trans4Goods. In: Future Security, Communications in Computer and Information Science. Berlin: Springer, pp. 25-28.

[38] Bohács, G. et al. (2012): Automatische visuelle Datensammlung aus Materialflusssystemen und ihre Anwendung in Simulationsmodellen. In: Logistics Journal nicht-referierte Veröffentlichungen.

[39] Koç, E. et al. (2013): Nutzung von Realdaten in Simulationsmodellen durch industrielle Bildverarbeitung. In: Dangelmaier et al. (Hrsg.) Simulation in Produktion und Logistik. Paderborn, HNI-Verlagsschriftenreihe 2013, pp. 349-359.

[40] Baechler, A. et al. (2016): The Development and Evaluation of an Assistance System for Manual Order Picking - Called Pick-by-Projection - with Employees with Cognitive Disabilities. In: Computers Helping People with Special Needs, Lecture Notes in Computer Science. Springer, pp. 321–328.

[41] Grzeszick, R. et al. (2016). Camera-assisted Pick-by-feel. Logistics Journal : Proceedings 2016.

[42] Funk, M. et al. (2016): motionEAP: An Overview of 4 Years of Combining Industrial Assembly with Augmented Reality for Industry 4.0. In: Proceedings of the 16th International Conference on Knowledge Technologies and Data-Driven Business, I-KNOW '16. ACM, New York, NY, USA.

[43] Hochstein, M. et al. (2017): Konsolidier-assistent - Assistenzsystem für manuelle Konsolidier- und Sortierprozesse in Distributionszentren. Logistics Journal: Proceedings 2017.

[44] Logivations (2018): Kamerabasierte Objekterkennung - Deep Machine Learning in Verbindung mit Computer Vision. url: http://www.logivations.com

[45] Qiu, X. et al. (2017): Hand Detection for Grab-and-Go Groceries (Course Project Reports). Stanford University.

[46] Motec GmbH (2016): Motec Kamera-Monitor-Systeme für Transport- und Kommunal-fahrzeuge - Mit Sicherheit weniger Unfälle, 2016. url: http://motec-cameras.com

[47] Salzer, S. (2018): Optisches Assistenzsystem zur sichere Handhabung palettierter Ware. url: http://www.nekos.exfa.de

[48] Lang, A. et al. (2017): Konzeption eines kamerabasierten Kollisionswarnsystems zur Prävention von Arbeitsunfällen an Gabelstaplern. In: Logistics Journal: Proceedings 2017.

[49] Trenkle, A. et al. (2013): FiFi – Steuerung eines FTF durch Gesten- und Personen-erkennung. In: Logistics Journal 2013.

[50] Overmeyer, L. et al. (2017): Intelligente Flurförderzeuge durch die Implementierung kognitiver Systeme. In: Handbuch Industrie 4.0 Bd.3, Springer Reference Technik. Berlin: Springer Vieweg, pp. 87-118.

[51] Thamer, H. et al. (2013): 3D-Bildverarbeitung für die automatische Entladung von Standardladungsträgern. In: Schenk, M. (Hrsg.): 18. Magdeburger Logistiktage - Sichere und nachhaltige Logistik, pp. 201-210.

[52] Weichert, F. et al. (2013): Automated Detection of Euro Pallet Loads by Interpreting PMD Camera Depth Images. Logist. Res. 6, 99–118.

[53] Eppner, C. et al. (2016): Lessons from the Amazon Picking Challenge: Four Aspects of Building Robotic Systems. Presented at the Robotics: Science and Systems XII.

[54] Uriarte, C. et al. (2016): Flexible Automatisierung logistischer Prozesse durch modulare Roboter. Logistics Journal : Proceedings 2016.

[55] Cao, L. et al. (2014): Peristaltik-Förderer - Integriertes Entlade- und Transportkonzept für Massenströme in der Paketlogistik: In: Schenk, M. (Hrsg.): Forschung vernetzen - Innovation beschleunigen: 15. Forschungs-kolloquium am Fraunhofer IFF, Magdeburg, pp. 7–13.

[56] Vatavu, A. et al. (2015): Modeling and Tracking of Dynamic Obstacles for Logistic Plants Using Omnidirectional Stereo Vision. In: 2015 IEEE/RSJ International Conference on Intelligent Robots and Systems (IROS), pp. 3552–3558.

[57] Hochstein, M. et al. (2016): Alternatives Linienführungssystem für autonome, fahrerlose Transportsysteme. In: Logistics Journal : Proceedings 2016.

[58] TU Chemnitz (n.d.): FOLLOWme - Intra-logistiksystem mit Fahrerlosen Transport-Systemen. url: https://followme-ils.com/

[59] Rieger, U. (2012): Visualisierung von Haftungsübergängen. url: http://www.hli-consulting.de

[60] Gebert, B. et al.(2012): Prozessvisualisierung auf Basis eines hybriden Sensorsystems. In: Go-3D 2012. Computergraphik für die Praxis, pp. 51–67.

APPLICATION OF PREDICTIVE ANALYSIS TO LOGISTIC PROCESSES IN GENERAL CARGO WAREHOUSES

Dipl.-Inf. Andreas Neubert
Logistik Competence Center
PKE Deutschland GmbH, Germany

1 Introduction
1.1 Area of application

General cargo can be defined as individualized, distinguishable goods that are handled individually [1]. General cargo warehouses are logistics nodes where general cargo is being handled. When goods arrive in or depart from warehouses, packages must be transported between the trucks and storage location. A low degree of automation exists e. g. in cross-docking through the use of underfloor conveyor chains or in high-bay warehouses through the use of driverless narrow aisle stackers [2]. Due to the different dimensions, shapes [3], weights and materials [4], general cargo is transported in warehouses by manually-operated industrial trucks such as forklifts and hand pallet trucks.

1.2 Problem definition

Manual activities are prone to human error. In the following, the arrival behavior of trucks in the general cargo warehouses, the storage location assignment and the forklift guidance system will be examined in greater detail for errors in logistic workflows. In part-load traffic trucks do not always arrive within the allocated time slot. Too late or too early arrival of the carrier has an impact on resources in the warehouse. If upon the receipt of goods, the package is unloaded at an incorrect storage location, this causes an unproductive search time for the package for subsequent loading. Putting down the package at an unfavorable storage location can result in long order picking routes for the subsequent goods issue. Once he has received the transport order, the forklift truck driver drives from his point of departure to the package and picks it up. He then transports the package to the place of destination and puts it down. From the place of destination, the forklift truck driver makes an empty run to the starting point, where he receives the next order. The empty runs back to the starting point are not resource-conserving. There is potential for optimization in planning and operational control.

1.3 Approach to solving the problem

For process visualization and documentation of the imputation of liability, tracking and tracing systems are used in general cargo warehouses. These systems consist of video technology, positioning systems and software for research. The systems are designed to document the current status in warehouses in 24/7 operation and archive the current status data in databases for the search for packages (tracking & tracing). These systems represent the state-of-the-art in general cargo warehouses.

In order to make predictions about future events, predictive analysis uses data-mining techniques [5]. A model is to be learned from the archived data using methods of data mining. The model should recognize patterns, e. g. for error situations. If current data is processed in the model in operational process control, the methods of predictive analysis can be used to detect error situations before they even occur in the warehouse. These methods are to be applied to logistic processes in the warehouse.

2 State-of-the-art

In addition to inventory management, a Warehouse Management System (WMS) comprises software tools for the control of the means of conveyance and a dock and yard management [6]. Figure 1 shows the core and additional functions of a WMS.

2.1 Dock/yard management

For strategic, medium-term and operational planning as well as control of different sub-areas of the occupancy planning of loading docks, scientific literature describes the methodological solutions of local search methods, branch-and-bound, genetic and evolutionary algorithms, scenario calculation, combinatorial auction and scheduling heuristics [7]. The study proposes a combined method consisting of branch-and-cut and an adapted column generation method for loading dock occupancy planning.

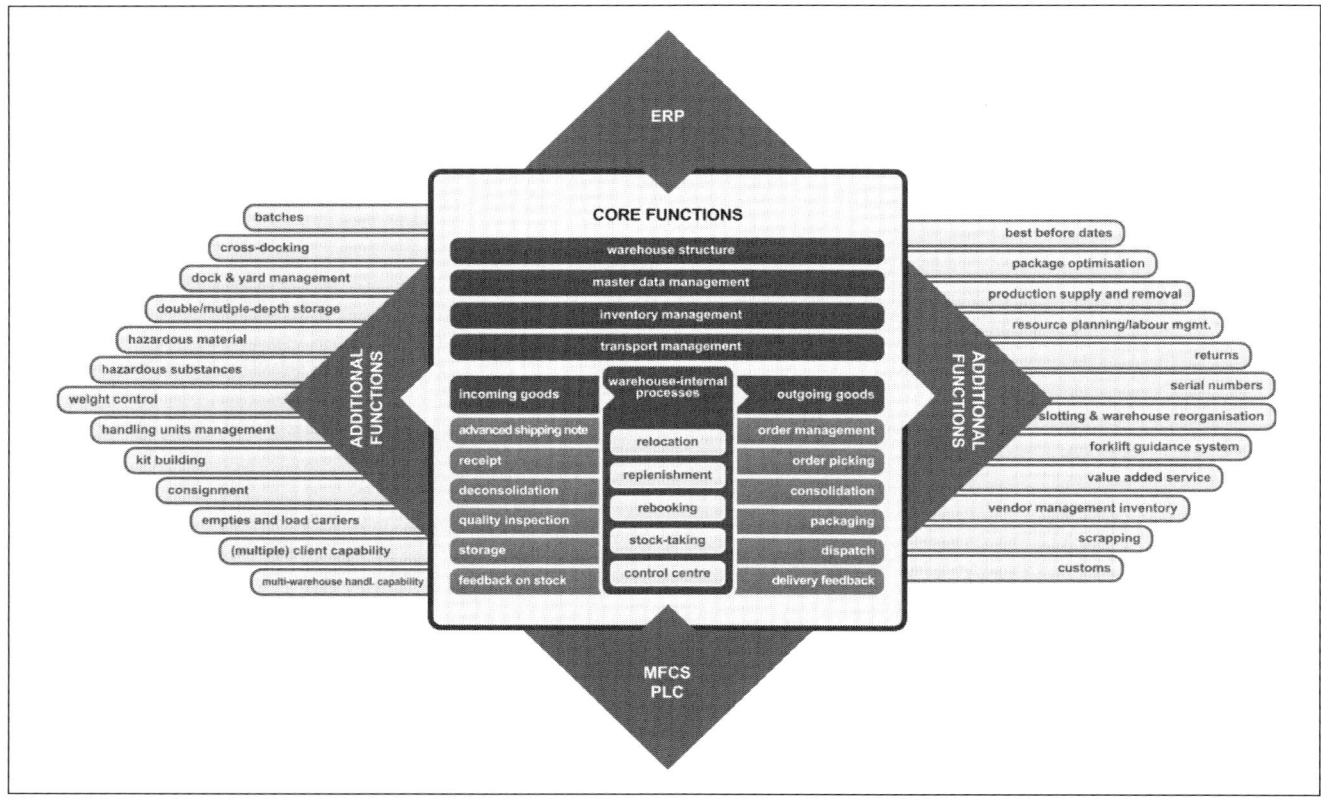

CORE FUNCTIONS

| warehouse structure |
| master data management |
| inventory management |
| transport management |

ERP

ADDITIONAL FUNCTIONS

batches
cross-docking
dock & yard management
double/mutiple-depth storage
hazardous material
hazardous substances
weight control
handling units management
kit building
consignment
empties and load carriers
(multiple) client capability
multi-warehouse handl. capability

best before dates
package optimisation
production supply and removal
resource planning/labour mgmt.
returns
serial numbers
slotting & warehouse reorganisation
forklift guidance system
value added service
vendor management inventory
scrapping
customs

incoming goods
advanced shipping note
receipt
deconsolidation
quality inspection
storage
feedback on stock

warehouse-internal processes
relocation
replenishment
rebooking
stock-taking
control centre

outgoing goods
order management
order picking
consolidation
packaging
dispatch
delivery feedback

MFCS
PLC

Figure 1: Core and additional functions of a WMS [8]

2.2 Warehouse management

Fixed location storage, free storage space allocation, zoning, across-aisle storage, part-family clustering, shortest transport route and anticipatory buffering are applied as strategies for storage space allocation upon entry into storage [9]. The strategies First-In-First-Out (FIFO), Last-In-First-Out (LIFO), quantity adjustment, preference granted to part-picked pallets, shortest transport route, minimizing aisle changes, tour-related and scheduled removal from storage, and precarriage are applied during removal operations [9]. The use of heuristics for forward-looking transportation of package units to the outbound goods area is also being suggested [10]. The study covers the combination of methods such as the relationships between storage, batching, zone picking and routing policies [11].

2.3 Forklift guidance system

There are different strategies for allocating orders to forklift trucks. The study examines the use of an agent-based auction procedure for assigning orders to forklift guidance systems [12]. This procedure is compared in simulations with priority-based strategies of First Come First Served (FCFS), Earliest Deadline First scheduling (EDF) and with various heuristics [12].

2.4 Data mining

Data mining involves the application of methodologies of data analysis and recognition

algorithms in order to detect patterns in the data [13]. Brandau analyzed the use of data mining in a cargo airport and laundrette [14]. Recognition algorithms include artificial neural networks and association rule learning. Azadnia studied association rule learning in combination with genetic algorithms for order batching with a view to minimizing delays [15]. Furthermore, association rule learning was evaluated for identifying zoning to assist picking operations in order to reduce picking routes and picking times [16]. Zhang examined the use of Hopfield neural networks to determine the shortest path for all orders in parallel order picking [17].

2.5 Assessment of the optimization methods

The procedures described above can be divided into
- mathematical methods,
- heuristics,
- simulation,
- genetic processes,
- auction procedures and
- techniques from the field of artificial intelligence.

Many of the proposed strategies neglect key logistic aspects and rules in general cargo warehouses [7]. The procedures were analyzed with planning tasks in mind.

The generated plan is then operationally executed by humans in warehouses, giving rise to error situations if packages are unloaded at the wrong storage location, for example. The methods described above fail to detect such errors.

It will be investigated in the following whether methods of predictive analysis are fit for identifying errors in operational processes beforehand in order to provide advance warning of manual unloading of packages at the wrong place of storage, thus saving costs incurred in unproductive search times.

3 Concept
3.1 Analytical methods

Analytical methods can be divided into descriptive, predictive and prescriptive techniques. Descriptive analysis answers the question what is happening right now. Positioning systems are used in this regard, for example, in order to determine the whereabouts of an item. While predictive analysis answers the question of what will happen, prescriptive analysis stipulates what should happen. It derives recommendations for action from descriptive and predictive analyses [18].

3.2 Suitability of predictive analysis for operational processes

A method is called for that takes all logistical workflows in general cargo warehouses into consideration and can be applied there.

It is first examined whether predictive analysis is suitable for process operations in general cargo warehouses. For this purpose, logistical processes in general cargo warehouses are compared with the application classes of predictive analysis.

Outliers may show up in the data generated in general cargo warehouses. Data for specific points in time may also be missing. The subsequent analysis of the data must be able to handle these data in such a way as to avoid skewing of results.

Predictive analysis provides clustering, classification, numerical prediction, association rule analysis, text mining and web mining [19] as application classes.

Available data can be classified into two categories (normal operation, error situation). Classification can be used to develop a model that assigns data to one of both categories. If the model is subsequently used to process current data from general cargo warehouses, either normal operations or error situations can be discerned. Future data developments can be predicted on the basis of numerical forecasts. Incorrect measurement values can thus be detected and eliminated and numerical

forecasting can be used to calculate missing data values. Relationships between data (attributes) are identified in association rule analysis, with if-then rules being established. Associative analysis enables warehouse staff to identify the conditions for error situations or the next package for removal from storage for example.

Several application classes of predictive analysis can be devoted to operational tasks in general cargo warehouses and methods of predictive analysis are, in principle, suitable for solving tasks in general cargo warehouses.

3.3 Sequence of predictive analysis

CRISP-DM standard is to be deployed as a methodology because CRSIP-DM is widely used [20]. CRISP-DM is a procedural model in which the phases business understanding, data understanding, data preparation, modeling, evaluation and data provision have to be completed [21].

Data preparation encompasses generating processable data from raw data, including the treatment of outliers and duplicates. Further information is acquired by applying data fusion to suitable data.

Historical data is used for modeling and the data volume is divided into training and test data. While training data is used to establish the model, test data is used to rate the quality of the trained model. Probability-based models, decision trees, rules, linear regression, logistic regression and neural networks are available for the tasks to be performed by modeling techniques. When feeding in the data, real-time data is applied to the model in order to be able to predict future events.

4 Realization
4.1 Determination of the current status

Operators of general cargo warehouses use cargo tracking systems in order to enable the tracking and tracing of packages in the warehouses. The consignment tracking systems consist of a video unit, positioning and software systems. Cameras are installed in the general cargo warehouse, with the images of all loading and unloading operations being stored in the video unit. So-called localization tags are attached to the barcode scanners. During the scan of a package barcode, the positions of scanners and the barcodes are stored in a database. Searches for barcodes can be conducted in the software system in order to display the corresponding video images of the event.

Cyber-physical systems from Industry 4.0 can also be integrated into general cargo warehouses. These devices record measurement data and transfer the data to a higher level.

Images taken by installed cameras can yield further benefits. Information about the current status of general cargo warehouses can be obtained by applying the algorithms of digital image processing. This information can subsequently be migrated to a higher level.

4.2 Using data from cyber-physical systems (CPS)

CPS consist of a sensor to record measurement data and a transmission module to transfer measurement data to another computer. When transmitting data, it is necessary to use a standard so that the recipient does not need to operate a large number of proprietary interfaces.

general cargo warehouses early on, movement data shall be considered in greater detail.

To this end, tracking and tracing systems can provide barcodes and tracking positions with corresponding time stamps. Once an industrial truck system has been located, the movements of packages can be recorded. If the loading condition of industrial trucks can be determined and reported, it is possible to ascertain whether an empty run is being conducted. It can be determined if and where a package is being stored, enabling warehouse staff to identify incorrect storage. When the status of the storage location has been recorded and reported, an early warning can be given of the storage area being

full, making it necessary to choose an alternative storage location for unloading.

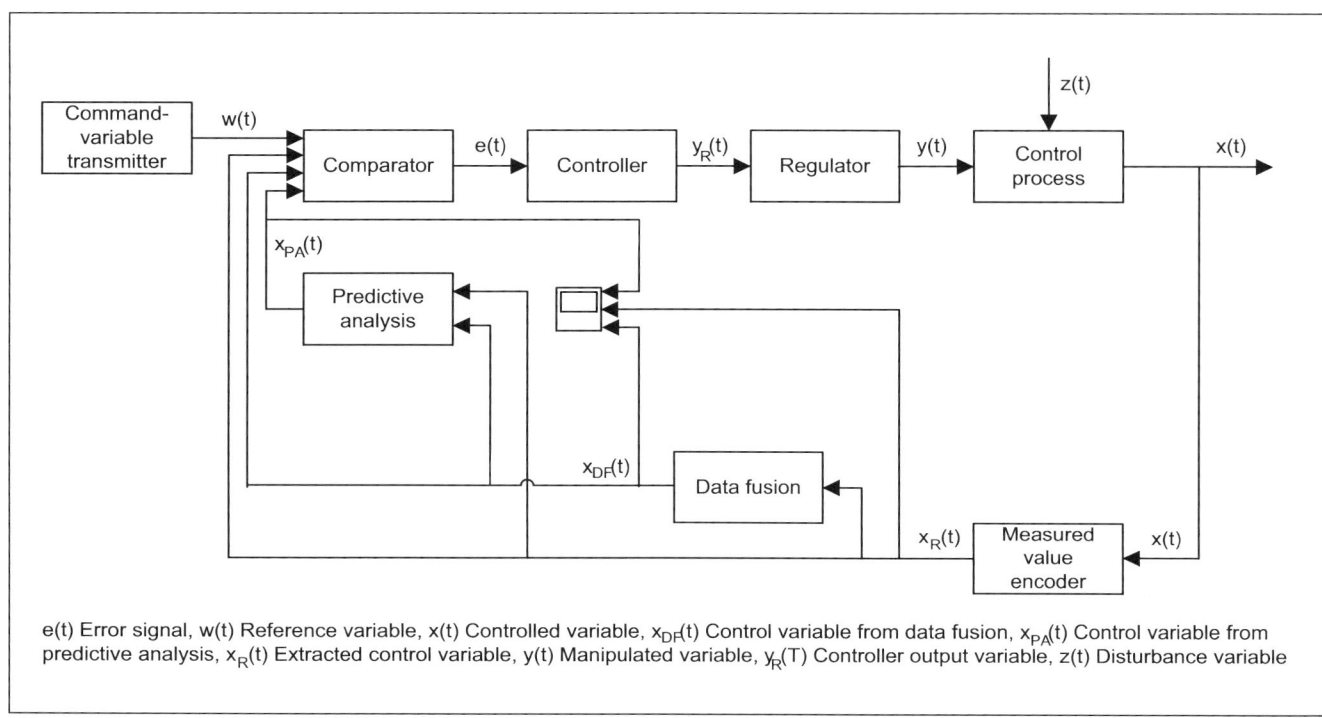

e(t) Error signal, w(t) Reference variable, x(t) Controlled variable, $x_{DF}(t)$ Control variable from data fusion, $x_{PA}(t)$ Control variable from predictive analysis, $x_R(t)$ Extracted control variable, y(t) Manipulated variable, $y_R(T)$ Controller output variable, z(t) Disturbance variable

Figure 2: Integration of predictive analysis into process control

Different transmission protocols have emerged in the Internet of Things. They differ according to both the type of communication (Push-Pull, Publish-Subscribe) and their overhead (header, type of transmission ...). Examples of communication protocols include HTTP, Extensible Messaging and Presence Protocol (XMPP), Constrained Application Protocol (CoAP), Advanced Message Queuing Protocol (AMQP) and Message Queuing Telemetry Transport (MQTT).One of these protocols should be used.

4.3 Description of the data to be used

Master data, stock data and movement data are available in general cargo warehouses. In order to detect error situations arising in transports in

4.4 Further data enable prediction of other events

If use is made of further data, planning tasks can be implemented alongside the operational workflow. In precarriage operations, packages containing outgoing goods to be loaded onto trucks soon are moved to the dispatch area in good time. Once the trucks have arrived in the goods issue area, they can be loaded faster. If the order and stock data as well as traffic jam reports are known, predictive analysis can be employed to forecast which truck will arrive at the area of goods issue and when precisely. In this way, the forthcoming order and thus the next packages can be better identified for precarriage to the outbound goods area.

4.5 Process analysis in general cargo warehouses

In process analysis the actual values that occur in a process are compared with the nominal values. Chapter 4.1. sets out how actual values can be obtained in logistical processes. The nominal values are defined as threshold values. Countermeasures must be taken if the threshold value is exceeded in the process.

In order to ensure better transparency in the workflows, the actual and nominal values are to be visualized to the dispatcher or warehouse clerk in a logistic monitoring cockpit providing an ergonomic user interface. This is designed to display logistic Key Process Indicators (KPI). Process analysis thus consists of
- several devices supplying up-to-date data on the current status of the logistic processes in general cargo warehouses,
- a process control program comparing real-time values with nominal values and
- a monitoring cockpit visualizing the logistic process with key process indicators.

4.6 Integration of predictive analysis into process analysis

The future values generated by predictive analysis assist the comparator of process control since the new control variable can thus be better identified. Figure 2 shows the planned sequence for the integration of predictive analysis in general cargo warehouses.

Consignment tracking systems document the current status in the warehouse in 24/7 operation and archive the current status data in databases for consignment search operations. If current data are processed in the trained model, the methods of predictive analysis can be used to detect error situations before they even occur in the warehouse. The methods of predictive analysis are to be used and reported for logistical processes in warehouses.

5 Conclusion

This article presented an optimization method for operational process control in manually-operated general cargo warehouses by applying predictive analysis.

The state-of-the-art of optimization approaches for logistical processes was determined, including an assessment of these approaches. The analysis covered dock/yard management, storage place assignment and the forklift guidance system. This was followed by an assessment of the suitability of predictive analysis as an optimizing approach for operational control in general cargo warehouses. The outcome of the study is that predictive analysis is suitable as an optimizing

approach for operational control with respect to the application classes classification, numerical prediction and association rule analysis. The article presented a concept for integrating predictive analysis into logistical processes in general cargo warehouses. Based on this, functions were defined in order to be able to use predictive analysis as an optimizing function in general cargo warehouses.

The next step is to realize the concept with the functions. The new optimization process is to be trained with historical data from a general cargo warehouse. After the model has been created, the optimizing method is to be applied to real-time data of operational control. The method is to be evaluated with respect to the recognition quality.

6 References

[1] Ten Hompel, M.; Heidenblut, V. (2011): Taschenlexikon Logistik: Abkürzungen, Definitionen und Erläuterungen der wichtigsten Begriffe aus Materialfluss und Logistik. Berlin, Heidelberg: Springer-Verlag, pp. 299-300.

[2] Ullrich, G. (2014): Fahrerlose Transportsysteme: Eine Fibel - mit Praxisanwendungen - zur Technik - für die Planung. Wiesbaden: Springer Vieweg Verlag, p. 96.

[3] Martin, H. (2016): Transport- und Lagerlogistik: Systematik, Planung, Einsatz und Wirtschaftlichkeit (E-Book). Wiesbaden: Springer Vieweg Verlag, pp. 59-60.

[4] Klaus, P.; Krieger, W.; Krupp, M. (2012): Gabler Lexikon Logistik: Management logistischer Netzwerke und Flüsse. Wiesbaden: Gabler Verlag, p. 545.

[5] Mishra, N.; Silakari, S. (2012): Predictive Analytics: A Survey, Trends, Applications, Opportunities & Challenges. (IJCSIT) International Journal of Computer Science and Information Technologies, Vol. 3 (3): 4434-4438.

[6] Bichler, K.; Krohn, R.; Philippi, P.; Schneidereit, F. (2017): Kompakt-Lexikon Logistik: 2.250 Begriffe nachschlagen, verstehen, anwenden. Wiesbaden: Springer Gabler Verlag, p. 241.

[7] Chmielewski, A. (2007): Entwicklung optimaler Torbelegungspläne in Stückgutspeditionsanlagen. Dortmund: Diss. Fakultät Maschinenbau, Universität Dortmund, pp. 42-43.

[8] Richtlinie VDI 3601:2015-09. (2015): Warehouse-Management-Systeme. Düsseldorf: Verein Deutscher Ingenieure (VDI), p. 30

[9] Ten Hompel, M.; Schmidt, Th. (2010): Warehouse Management: Organisation und Steuerung von Lager- und Kommissioniersystemen. Berlin, Heidelberg: Springer-Verlag, pp. 31-33.

[10] Jahani, P. (2016): Dynamic warehouse optimization using predictive analytics, Electronic Theses and Dissertations, Paper 2582. University of Louisville, p. 167

[11] Van Gils, T.; Braekers, K.; Ramaekers, K.; Depaire, B.; Caris, A. (2016): Improving Order Picking Efficiency by Analyzing Combinations of Storage, Batching, Zoning, and Routing Policies, A. Paias et. al. (Eds): ICCL 2016, LNCS 9855, Springer International Publishing Switzerland, pp. 427-442.

[12] Mirlach, M.; Günthner, W.; Ulbrich, A.; Beckhaus, K. (2013): Auftragszuteilungs-verfahren für Staplerleitsysteme, 17. Flurförderzeugtagung 2013. Düsseldorf: VDI Verlag, pp. 67-78.

[13] Fayyad, U.; Piatetsky-Shapiro, G.; Smyth, P. (1996): From Data Mining to Knowledge Discovery in Databases. AI Magazine Volume 17 Number 3: 37-54.

[14] Brandau, A. (2015): Ganzheitliches Konzept zur Modellierung und Analyse von Zustandsdaten logistischer Objekte. Magdeburg: Diss: Institut für Logistik und Materialflusstechnik. Otto-von-Guericke Universität Magdeburg, p. 115

[15] Azadnia, A. H.; Taheri, S.; Ghadimi, P.; Saman, M. Z. M.; Wong, K. Y. (2013): Order Batching in Warehouses by Minimizing Total Tardiness: A Hybrid Approach of Weighted Association Rule Mining and Genetic Algorithms. The Scientific World Journal, Volume 2013, Article ID 246578, 13 pages.

[16] Chuang, Y.; Chia, S.; Wong, J. (2014): Enhancing Order-picking Efficiency through Data Mining and Assignment Approaches. WSEAS TRANSACTIONS on BUSINESS and ECONOMICS, E-ISSN: 2224-2899, Volume 11: 52-64

[17] Zhang, H.; Zhu, J.; Zhou, L. (2015): Study on Order Batching Model Design Based on Hopfield Neural Network. Science Journal of Business and Management. Vol. 3, No. 2: 60-64.

[18] Souza, G. C. (2014): Supply chain analytics. The Journal of the Kelley School of Business, Indiana University, Business Horizons 57: 596-605

[19] Cleve, J.; Lämmel, U. (2016): Data Mining. Berlin, Boston: De Gruyter Oldenbourg, pp. 57-67.

[20] Brandau, A. (2015): Ganzheitliches Konzept zur Modellierung und Analyse von Zustandsdaten logistischer Objekte. Magdeburg: Diss: Institut für Logistik und Materialflusstechnik. Otto-von-Guericke Universität Magdeburg, p. 72

[21] Chapman, P.; Clinton, J.; Kerber, R.; Khabaza, Th.; Reinartz, Th.; Shearer, C.; Wirth, R. (2000): CRISP-DM 1.0. USA, Denmark, Germany, The Netherlands: CRISP-DM consortium, pp. 10-12.

CONTINUOUS IMPROVEMENT OF LEAN PROCESSES WITH INDUSTRY 4.0 TECHNOLOGIES

M. Sc. Sven Rittberger
Technology Centre for Production and Logistics Systems (PULS)
Landshut University of Applied Sciences, Germany

Prof. Dr. Markus Schneider
Technology Centre for Production and Logistics Systems (PULS)
Landshut University of Applied Sciences, Germany

1 Introduction

In an increasingly globalized and digitized economy, manufacturing companies are facing growing competition. Successful are those that produce and supply the products requested by the customer in higher quality, faster and more cost-effectively than the competition. In order to achieve this, it is necessary to constantly increase the efficiency of production [1].

In research and practice, a well-established approach to planning efficient production and logistics processes is the design of lean, waste-free value chains according to Lean Production [2]. A central principle is the Continuous Improvement Process (CIP). By small daily improvements, employees at all organizational levels systematically eliminate activities in their immediate work environment that bind resources from the customer's perspective and do not add value to the product [3].

Another approach to increasing the efficiency of production and logistics processes is the implementation of Industry 4.0 solutions. By implementing and interconnecting technologies, information flows to employees and between processes and systems of an organization are systematically designed and automated [4].

While the focus of Lean is on an optimal interaction between human and organization, Industry 4.0 outlines how the interrelations between human and technology as well as organization and technology can be structured in the best possible way. A comprehensive method for analyzing these relationships is the concept of human-technology-organization (HTO) [5]. The three dimensions including their interfaces are shown in Figure 1.

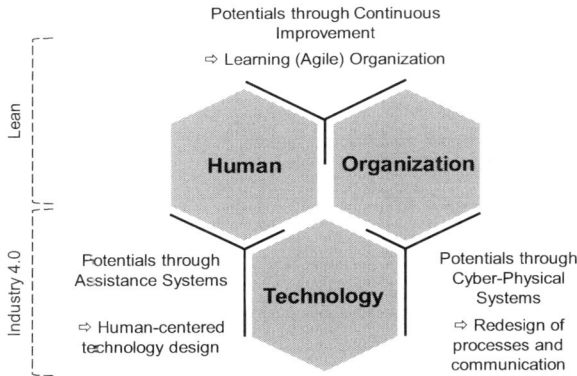

Figure 1: Potentials of Lean and Industry 4.0 structured by the HTO concept (own representation following [5])

2 Optimal design of the human-organization intersection through CIP

Although CIP is often associated with the Toyota production system, it has its origins in the Plan-Do-Check-Act Cycle ("PDCA Cycle") published by W. Edwards Deming. In four phases, it describes a systematic approach for the creative solution of problems (see Figure 2) [3].

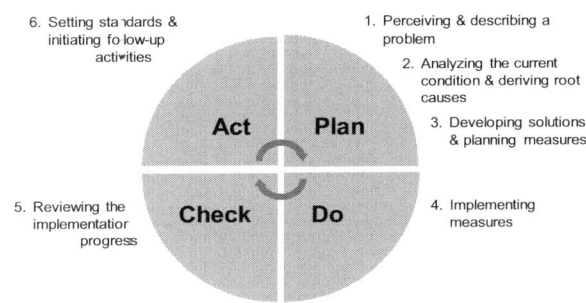

Figure 2: Phases of the PDCA Cycle (own representation following [3])

The improvement cycle starts with the Plan phase, in which an occurring problem is perceived and described by the concerned employee on site. In the second step, influencing factors that have led to the problem are analyzed. In small teams, the actual situation is observed on site and systematically questioned. In this process, a mostly complex problem is broken down step by step until the cause can be narrowed down. Once the concerned employees have defined the problem and identified the root cause, joint solutions for a sustainable solution to the problem are developed and appropriate measures are planned [3].

CIP is based on a culture of collaborative problem solving. It focuses on employees of all company levels as creative and adaptive problem solvers in all optimization activities [6]. Through daily improvements, they develop a deep understanding of processes and experience in teams that empowers them not only to treat the symptoms of inefficient processes, but to systematically identify their hidden root causes and eliminate them with creative solutions [7].

In the second phase, the Do phase, measures that were defined initially are implemented. These improvements are done in many small steps instead of radical but time-consuming leaps. By experimenting in short cycles, employees are able to observe the direct effects of their changes and can immediately learn from them [8].

The Check Phase aims to review the implementation progress. For this purpose, the success of implemented measures is compared against the target condition that has been defined in the Plan phase. Transparency of the condition of a process and its implementation progress is only possible by measuring and openly visualizing objectives and results.

Once a solution is implemented, the act phase starts and thus the reflection on the results achieved. New organizational process standards are set, and the knowledge gained is the starting point for a new improvement cycle at a higher level [3].

By constantly repeating the PDCA loop, all processes within an organization become increasingly efficient and attained standards can be maintained. This clarifies one of the Parkinson laws: "[...] once an organization forms its structure it starts to move backward" [8]. Through continuous improvement, employees find themselves in a continuous reflection and learning process that leads to a continuous and agile development of the entire organization. If this goal is achieved, it is referred to as a learning organization.

Although an organizationally well-established CIP offers many advantages, there is a considerable deficit between theory and practice. During the practical execution, companies encounter obstacles that lead to a CIP being implemented only partially [9]. These obstacles are outlined in the following.

3 Obstacles during the practical execution of the CIP

3.1 Obstacles during the Plan phase

Today's value chains present production employees with the challenge that processes become increasingly complex and intransparent [10]. If a deviation from the target and actual condition of a process occurs, employees do not notice this during day-to-day operations or do not perceive it as a problem. Due to a shortage of personnel capacities, some improvement loops already fail in the detection of process deviations [3].

If employees detect problems, this happens at the earliest on occurrence, but usually with a certain latency. In CIP, problems are analyzed and solved post-mortem, but they cannot be prevented in advance [11].

Furthermore, increasingly complex and intransparent processes lead to cause-and-effect relationships that become increasingly difficult to understand [8]. If a problem arises in practice, employees often take intuitive and immediate countermeasures. Since a systematic analysis of the current state and the derivation of root causes are neglected, they lack metrics, measures and indicators that are needed to understand complex problems. Misinterpretations and misinterpretations of problems occur, which lead to the fact that only symptoms are eliminated, or problems are even worsened [3].

In accordance with the management principle "If you can't measure it, you can't manage it." [12] a dataset is the basis for deriving correlations between problems and their underlying root causes, as well as for evaluating improvement progress in the Check Phase [3]. However, the manual collection of figures, data and facts requires a lot of effort for employees. This makes the Plan phase the most extensive part of CIP, and also requires a profound experiential knowledge of the employees, which can only be developed over several years of practice [7]. This ties up personnel resources that are used for operational activities in practice [13].

3.2 Obstacles during the Do phase

Not only the planning of improvement activities, but also their implementation is executed in a structured CIP together with the operative employees on site. Only they possess the necessary experience to implement their own solutions creatively and with the simplest means [8]. In practice, however, there is a lack of capacity for this purpose [14]. Improvement measures are subordinated to day-to-day business and are often only implemented in accordance with the necessary minimum requirements [3].

3.3 Obstacles during the Check phase

A mechanism for implementing solutions insufficiently is embedded in the Check phase of CIP. Here the team reviews whether the initially defined targets have been achieved. Employees independently collect and visualize process data in defined cycles. For the greatest learning effect possible, these activities are carried out manually by each employee. Hereby, companies encounter an additional obstacle, because in addition to the analysis of the current condition, the derivation of root causes and the implementation of countermeasures, further capacities are demanded from production employees for the comparison between current and target condition. For this reason, an irregular or insufficient progress review can be observed in practice [3].

3.4 Obstacles during the Act phase

A phase being largely neglected in practice is the Act phase. Responsible persons of the CIP face organizational obstacles, especially when introducing company-wide standards. Although many companies establish a formal knowledge management system, the specialization of corporate functions reduces the effectiveness of knowledge transfer in practice. Process standards are filed in the databases of the various departments without being used for subsequent improvements. In the departments, knowledge management only exists in the form of implicit experience knowledge of the employees [3]. In order to achieve the CIP goal of a learning organization, knowledge has to be communicated across organizational silos [15].

4 Potentials for enhancing the CIP by technology

The continuous improvement process enables an optimal interaction of the human-organization intersection. The PDCA cycle is traditionally executed with minimal technical assistance so that employees can perform and understand each improvement step themselves [3]. The latest information and communication technologies offer opportunities to support people in their activities. According to the vision of Industry 4.0, the end-to-

end digitization of manufacturing offers potentials to increase the productivity of employees and organizational processes [4]. In the following, it is described which potentials the design of the intersections human-technology and organization-technology offer to overcome the mentioned obstacles during the practical execution of the CIP. An overview is shown in Figure 3.

4.1 Potentials through the design of the intersection human-technology

A core relationship of the HTO concept is the intersection of human and technology. This socio-technical approach places the employee at the center when implementing technologies. The goal is to support and relieve humans in their activities. This can be achieved either cognitively by supplying employees with relevant information automatically, or physically during physical activities.

Regarding the CIP, people can primarily be supported with the former. Industry 4.0 offers solutions in the form of assistance systems that support employees in monitoring conditions or making complex decisions [5].

4.1.1 Potential 1: Condition monitoring of the occurrence and resolution of problems

In the CIP's Plan Phase, employees are faced with the challenge of detecting process deviations or problems respectively. In addition to personal experience, the visualization of process standards (e.g. floor markings) helps to identify process anomalies. Industry 4.0 also offers further solutions for visualizing process deviations. By connecting physical production objects and virtual information systems, process data can be collected automatically and in real time. If the parameters of a process exceed defined limits in such a digitized production environment, it indicates problems and can be, in conjunction with relevant background data, displayed to the employee at his or her workplace without any delay. In addition, the progress of improvement measures no longer has to be checked manually,

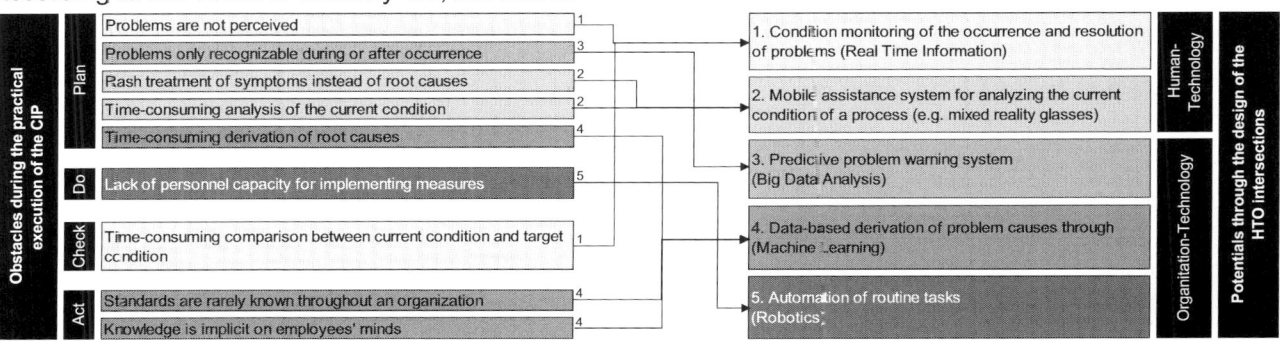

Figure 3: Potentials for enhancing the CIP with Industry 4.0 technologies

but a comparison of the current and target values is carried out continuously and automatically [4, 6]. The design of the intersection between human and technology provides a continuous transparency of the real-time condition of a production environment and arising problems.

4.1.2 Potential 2: Assistance system for analyzing the current condition of a process

Providing real-time data to employees reduces the barrier to rashly implementing intuitive countermeasures and thus reducing the risk of just correcting the symptoms of a problem instead of its root cause. This allows people to query specific process information that they need to understand a problem and to narrow down its root cause. It is important to ensure that employees continue to analyze and understand problems on the shop floor. If analyses are performed exclusively data-driven on a screen, the human potential as a creative and adaptive problem solver remains largely unused [16]. Mobile assistance systems in the form of portable devices, such as tablet computers or mixed reality glasses, facilitate this. The employee is supported in the analysis of process steps with relevant data, which increases transparency and thus the understanding of the problem. If a problem can be solved on site, the employee is also able to initiate the implementation of an immediate measure directly via the mobile device. Increasing the inventory level of a supermarket is such an example [17]. With the design of the human-technology intersection it is possible for employees to analyze the current situation more efficiently and effectively.

4.2 Potentials through the design of the intersection organization-technology

The third relationship of the HTO concept comprises the intersection between organization and technology. Here again technology is assigned a supporting function. With the help of Industry 4.0 solutions it is possible to design processes and communication of an organization under completely new conditions. Above all, merging the physical and virtual worlds into cyber-physical systems allows new and innovative design possibilities for the CIP [4, 5].

4.2.1 Potential 3: Predictive problem warning system

If the condition monitoring described above is implemented, the challenge remains that problems can only be detected when they occur and therefore cannot be prevented in advance. The basis of condition monitoring is a very large amount of structured and unstructured production data. If combined, not only known correlations emerge, but also hidden patterns and trends that

predict process anomalies and thus problems [11]. Summarized under the term big data analysis, this information helps employees to identify complex cause-and-effect correlations more quickly and proactively initiate countermeasures. Designing the intersection between organization and technology leads to fewer interruptions and thus to more stable processes [18].

4.2.2 Potential 4: Data-based derivation of problem causes

If an improvement measure is successful, it is filed as a standard in databases during the act phase. Although these standards include important information about problems, causes and effective measures, they are rarely known throughout an organization. For new problems, employees draw on the limited experience of their group and often work on problems that have already been solved by other teams [3]. These communication obstacles can be overcome through a cross-functional design of information flows. If there is a process deviation or a problem respectively, employees are automatically informed about known root causes. These cause-problem relationships originate from a central database that is updated in the Act phase of each improvement cycle. With every successfully completed improvement, a new data set is created, which is used to train a learning algorithm. The described method of machine learning corresponds to the learning process of humans during the CIP and enables an organization-wide distribution of experiential knowledge about problem root causes [18]. By designing the organization-technology intersection, the effort for a time-consuming derivation of root causes can be significantly reduced.

4.2.3 Potential 5: Automation of routine tasks

Even if the presented approaches remove obstacles during the practical execution of the CIP in the plan, check and act phases, employees lack the capacity to implement improvement measures during the Do phase. While in today's multi-variant operations, employees are mainly busy assembling products, machines are increasingly taking over these activities. From the perspective of CIP, the ongoing automation associated with Industry 4.0 relieves employees of executing repetitive routine tasks and allows them to use the freed capacities for improvement activities. By designing the intersection between organization and technology, physical processes can be automated, and humans with their creativity and experience can contribute to the continuous increase of value added through process improvements [19].

5   Conclusion and outlook

This paper presents a joint approach to improve production and logistic processes with Lean Management and Industry 4.0. For this purpose, the concept of human-technology-organization (HTO) is used to demonstrate and structure resulting potentials.

Lean foremost enables an optimal interaction between humans and their organization which leads to a learning and agile organization through Continuous Improvement. Industry 4.0 Technology offer additional solutions for enhancing process improvements through assistance systems and cyber-physical systems. Assistance systems are technologies that support and relieve humans in monitoring or analyzing the condition of a process. They provide real-time information on mobile devices such as tablet computers and mixed reality glasses. Cyber-physical systems, on the other hand, offer completely new opportunities to redesign processes and communication within an organization and its peers. Technologies such as big data, machine learning and robotics facilitate process improvements.

Further research needs to analyze where the presented technological solutions are already used to enhance the continuous improvement process.

6   References

[1]   Schenk, M.; Wirth, S.; Müller, E. (2014): Fabrikplanung und Fabrikbetrieb – Methoden für die wandlungsfähige, vernetzte und ressourceneffiziente Fabrik. Berlin, Heidelberg: Springer, pp. 17-18, pp. 420-421.

[2]   Schneider, M. (2016): Lean Factory Design – Gestaltungsprinzipien für die perfekte Produktion und Logistik. Munich: Carl Hanser, pp. 63-68.

[3]   Kostka, C.; Kostka, S. (2017): Der kontinuierliche Verbesserungsprozess – Methoden des KVP. Munich: Carl Hanser, pp. 5-50, p. 60.

[4]   Kagermann, H.; Helbig, J.; Hellinger, A.; Wahlster, W. (2013): Umsetzungsempfehlungen für das Zukunftsprojekt Industrie 4.0 – Abschlussbericht des Arbeitskreises Industrie 4.0. Berlin, Frankfurt/Main: acatech – Deutsche Akademie der Technikwissenschaften e. V., pp. 23-26.

[5]   Reinhart, G. (2017): Handbuch Industrie 4.0 – Geschäftsmodelle, Prozesse, Technik. Munich: Carl Hanser, pp. 54-59.

[6]   Dickmann, P. (2015): Schlanker Materialfluss – Mit Lean Production, Kanban und Innovationen. Berlin: Springer Vieweg (VDI-Buch), pp. 26-27.

[7]   Rother, M. (2013): Die Kata des Weltmarktführers – Toyotas Erfolgsmethoden. Frankfurt, New York: Campus, pp. 32-33, pp. 220-221.

[8]   Imai, M. (1997). Gemba Kaizen – A Commonsense, Low-Cost Approach to Management. New York: McGraw-Hill, pp. 1-26.

[9]   Meister, M.; Metternich, J.; Batz, S. (2017): Reifegradmodell für den systematischen Problemlösungsprozess. ZWF 112: 848–851.

[10] Bennett, N.; Lemoine, G. (2014): What VUCA Really Means for You. Harvard Business Review. Retrieved 05 Janurary 2018, from https://hbr.org/2014/01/what-vuca-really-means-for-you.

[11] Göger, C. (2015): Advanced Manufacturing Analytics – Datengetriebene Optimierung von Fertigungsprozessen. Lohmar: Josef Eul, pp. 9-13.

[12] Kaplan, R.; Norton, D. (1996): The Balanced Scorecard – Translating Strategy into Action. Boston: Harvard Business School Press, pp. 21.

[13] Smith, G. (1989): Defining Managerial Problems – A Framework for Prescriptive Theorizing. Management Science 35:963-981.

[14] Bosch Software Innovations GmbH (2015): Industry 4.0 Market Study – Demand for Connected Software Solutions. Berlin: Bosch Software Innovations GmbH, pp. 19-21.

[15] Lüthy, W. (2002): Wissensmanagement-Praxis – Einführung, Handlungsfelder und Fallbeispiele. Zürich: vdf, pp. 7-28.

[16] Liker, J. (2003): The Toyota Way – 14 Management Principles from the World's Greatest Manufacturer. New York: McGraw-Hill, pp. 149-158.

[17] Coia, A. (2017): Robert Bosch 4.0 – Keeping a close eye on technology. Retrieved 03 Janurary 2018, from https://automotivelogistics.media/intelligence/robert-bosch-keeping-close-eye-technology.

[18] Schuh, G., Anderl, R., Gausemeier J., ten Hompel, M., Wahlster, W. (2017): Industrie 4.0 Maturity Index – Die digitale Transformation von Unternehmen gestalten (acatech STUDIE), Munich: Herbert Utz Verlag 2017, pp. 10-18.

[19] Spath, D.; Ganschar, O.; Gerlach, S.; Hämmerle, M.; Krause, T.; Schlund, S. (2013): Produktionsarbeit der Zukunft – Industrie 4.0. Stuttgart: Fraunhofer Verlag, pp. 50-55.

DESIGN ASPECTS OF LAST MILE LOGISTICS SOLUTIONS

János Juhász, PhD student
Institute of Logistics
University of Miskolc, Hungary

Tamás Bányai, PhD
Institute of Logistics
University of Miskolc, Hungary

1 Abstract

Today's competitive operation of supply chain management is one of the highest priority parts of logistics to ensure maximum utilization of resources. Aim of supply chain management (SCM) is serving the customers' demands, the manufacturing and a service processes.
Also the global market and increased product variety has forced the manufacturing processes and their logistics service provider to be more effective.
Vehicle Routing Problems (VRP) represents a difficult task of transportation. Our model will determine the optimal route between transportation tasks, running transportation tasks to additional demands for small number of routes, which take into consideration the transportation operation costs and capacity occupancy. Industry 4.0 provides different tools to implement latest principles and one of our tools will be a vehicle routing planning application. The main goal of this article is to design a general model, which could describe supply chain problems for small number of routes.

Keywords: Vehicle Routing Problem, last mile, logistics, design, supply chain.

2 Introduction

The supply chain management (SCM) development is a key challenge for logistics area, especially in production area, health care and in related logistics services. It means these systems need to serve society' demands. The participants require access to get up-to-date information to find the fastest way, which can fulfil the requirements of any transportation tasks. To create a efficient system, where all participants can communicate and serve each other's demands need network resources management.
A Vehicle Routing Problem (VRP) can be found in today' transportation systems and logistics' applications. VRP become more and more important in the field of transportation planning and the logistics applications. At the same time, the efficiency, capacity and timeliness become more and more important. VRP is an NP-hard optimization problem. VRP is a marked problem for the logistics transportation in today's research. It contains the transport vehicles diversity, the complexity of networking partners [1]. This cooperation aims to enable highly efficient and sustainable supply chain solutions based on intelligent transportation systems and geographical information systems using up-to-date technologies.
In the modern world, we need to be competitive, and use up-to-date methods and keep steps with industrial trends, which take into consideration financial especially transportation-related costs and environmental load.
Over the last few years, researchers have focused on design last mile methods, the most relevant scientific results need to be summarized before elaborate integrated models, algorithms and solutions.
The aim of this paper is to identify challenges of vehicle routing planning in the last mile logistics solutions from the aspects of Industry 4.0 smart solutions. The Industry 4.0 provides different tools to implement latest principles. One of these tools can be the vehicle routing planning application. It is a possible way to determine optimal path for small number of routes, and it will support further studies to optimize large number of routes.

This paper is organized as follows: Section 2 presents a literature review, which summarizes the research results related to last mile aspects of supply chain operation. Section 3 describes a generate model, which focuses on operation cost. Section 4 demonstrates the numerical analysis of the model. Conclusions and future research directions are discussed in Section 5.

3 Literature review

Since our study embraces two related streams, called last mile solutions in logistics and VRP, we can provide a brief review on each stream before to present our approach.

3.1 Last mile solutions in logistics

As we can find in the literature, logistics has undergone many scientific changes in the last few years. Modern and innovative strategies, where solutions enable implementation of smart supply chain management. These new aspects allow handling in a more effective way the last mile delivery tasks, which focusing on cost of operation [2].

Most of studies analyzed distribution network problems, and their solutions to allow more accessible and equitable distribution of relief supplies [3]. The freight transportation has become a key challenge in the development of countries, and the transportation tasks are supported by decision support systems [4]. Delivery service providers should concentrate on their marketing force and customize their services for consumer groups who have specific individual characteristics, such as optimism and innovation [5].

To serve out costumer demands are difficult task for urban areas. New methods and NP-hard problems optimization are required. To solve transportation task and its problems, algorithms and models are proposed. Hybrid multi-population genetic algorithm is designed to generate initial solutions [6]. The aim is to reduce inefficient vehicle movement, using an agent-based model [7] and maximize profit, consisting of the revenue from users [8].

Other algorithms also can solve problems, to improve the quality and reduce transportation costs. The transportation tasks and their problems also mentioned as traveling salesman problem [9]. Improved e-commerce efficiency for order handling allows logistic service providers to better align strategically with online retailers [10].

3.2 Vehicle Routing Problem

Vehicle routing problem is important research area of operational studies. Several international companies like Amazon, and UPS shows interest in VRP, to optimize transportation tasks fulfill processes [11].

The VRP solutions are used in application of automated vehicles, which designed to execute all delivery tasks, and considering capacity occupancy [12].

The articles that addressed the I4.0 application is focusing on new route planning to identify the logistic aspects from design and operation point of view of supply chain. It was found that VRP is important for vehicle routing planning.

4 General model

Within the frame of this chapter a general model of assignment of transportation resources and transportation tasks is introduced. This study focuses on vehicle routing problems of network activities and services of companies how can fulfil additional new tasks using initial routes.

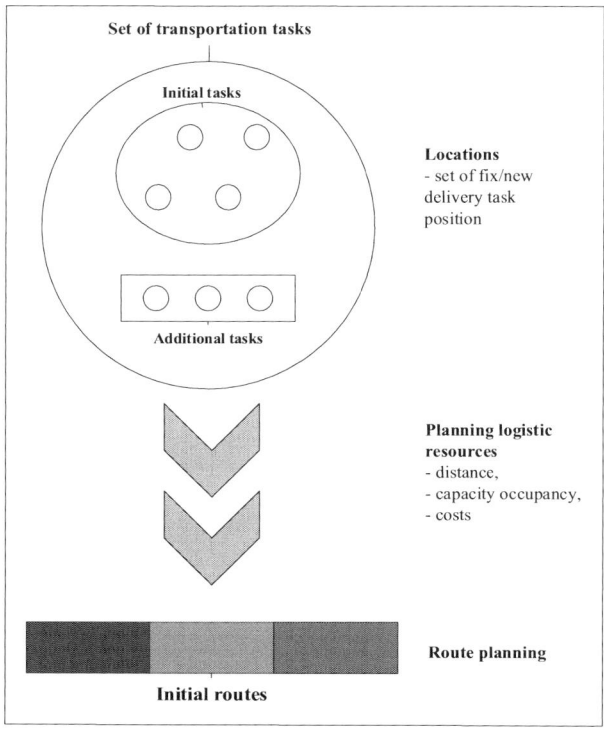

Figure 1: Route planning theoretical model

The model represents the supply chain process, where geographical data and other information are available through I. 4.0. Cloud. The transportation tasks including additional tasks of the system can be assigned to available logistic resources see Figure 1.

The model framework of route planning between objects (fix and additional) each route to fulfil set of transportation tasks. The decision variable defines the decision to be made. In this theoretical scenario, the following decisions must be made: (1) assignment of locations to transportation tasks, (2) assignment of transportation tasks to logistic resources, (3) find the bee line. With this mind we can define the following decision variable:

— $x_{n,m}$ is the assignment matrix of locations, and logistic resources, where n=1…i, and m=1…l.

The objective function of the problem describes the minimization of the transportation routes cost, which is the operation cost.

$$c = \sum_{n=1}^{i} \sum_{m=1}^{l} x_{n,m} \left(c_{n,m}^{R}(l_t(D_T, \text{CAP}_0), l_r, c_{n,m}) \right) \to min. \quad (1)$$

where

- $c_{n,m}^P$ is the operation costs of route, where n^{th} location of transportation task is assigned to m^{th} logistic resources,
- l_t is the position of transportation tasks, which depends on D_T, the optimal way of objects and CAP_o, the capacity occupancy,
- D_T is the shortest way between transportation tasks on each routes,
- l_r is the position of mth logistic resources,
- $c_{n,m}$ is the specific cost of n-m relation, where n^{th} location of transportation task is assigned to m^{th} logistic resources.

To find the optimal way, as shortest way between transportation tasks on each route can be calculated by bee line measuring.
In general, there are three options to measuring distance between two points: bee line-, track- and road distance measuring. We used air line distance to find optimal way, usually formalized bee line measuring. The location coordinates are known.
We consider the capacity of vehicles, because each vehicle has different occupancy and different transportation tasks with package request. With this mind we can define the capacity occupancy:

$$Cap_o = \sum_{n=1}^{i} \sum_{m=1}^{l} \left(Cap_q^R (\max) \pm Cap_q^{Tt} \right) + Cap_q^{NT} \qquad (2)$$

where

- Cap_q^R is the maximum capacity of vehicle route,
- $\pm Cap_q^{Tt}$ is the quantity of each initial transportation tasks, depends on task request (delivery/pick up),
- Cap_q^{NT} is the quantity of the new task.

Some constraints are introduced for boundary conditions, and this makes the methodology more complex. The constraints are defined as the following:

The location coordinates are not defined, and generated with random values. We applied one of Microsoft Excel function to generate the elements of coordinates and it returns a new random number each time your spreadsheet recalculates. With this mind we know the syntax of this function as follows:
    =RANDBETWEEN(bottom, top) function returns a random integer between bottom and top value.
- The quantity of each initial transportation tasks is already pre-specified.
- The quantity of additional transportation task is generated with the instructed function of Microsoft Excel. The bottom and top value range are known.
- The maximum capacity of vehicle routes (Cap_q^R) are defined.
- Transportation tasks must be done during working hours, from 8:00 am to 4 pm.

5   Scenario analysis

Within the frame of this chapter, scenario will be presented to demonstrate the efficiency of route planning in aspect of last mile logistic solution of supply chain services.
The above-described model makes it possible to define an optimal solution of route planning with different assignments.
Figure 1 describes the route planning process, where Industry 4.0 tool was implemented.
There are two cases to test the functioning and the efficiency of the method we should look at some examples. In the first case, the parameters of the first route are described in Table 1.

Initial transportation task	Time	Loc$_x$	Loc$_y$	Quantity of task [unit]
1.	8:15	5	9	20
2.	9:05	10	18	13
3.	9:33	15	33	11
4.	9:55	22	35	13
5.	10:20	25	43	8
6.	10:40	36	55	10
7.	11:16	50	60	14
8.	13:20	55	63	9
9.	14:40	60	44	13
10.	15:40	70	56	2

Table 1: Route parameters of the first vehicle

Maximum capacity of the vehicle (each route) is given and quantity in units. We assigned time to fulfil demands; location coordinates of transportation tasks and quantity of tasks. The parameters of the second route are described in Table 2. The vehicle's capacity impact is not analyzed in this case.

Initial transportation task	Time	Loc$_x$	Loc$_y$	Quantity of task [unit]
1.	8:00	3	12	5
2.	9:30	20	18	14
3.	10:20	25	25	1
4.	11:10	35	36	9
5.	12:00	44	40	4
6.	12:20	49	47	14
7.	13:40	58	49	7
8.	14:10	63	53	1
9.	14:50	66	30	9
10.	15:20	75	55	6

Table 2: Route parameters of the second vehicle

As Table 1 and Table 2 shows, the parameters can provide for n, each vehicle route. If the location of the additional task is 11 (38; 43), then the shortest way between transportation tasks on each route can be calculated with bee line measuring. The results are depicted in Figure 2.

1. route	DT1 [unit]	2. route	DT2 [unit]
1.	47	1.	47
2.	38	2.	31
3.	25	3.	22
4.	18	4.	8
5.	13	5.	7
6.	12	6.	12
7.	21	7.	21
8.	26	8.	27
9.	22	9.	31
10.	35	10.	39

Figure 2: D$_T$ results

In the case of the scenario the shortest, optimal way can be found in the vehicle route problem. It is 7 unit. Figure 3 shows the 5th initial task is the nearest to additional task, after then the new task can be fulfilled.

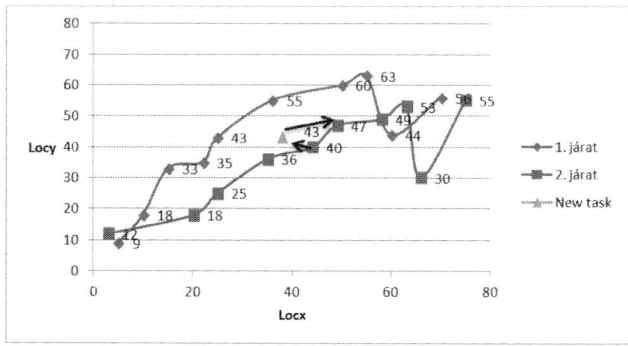

Figure 3: Solution of first case

The nearest transportation tasks can be taken into consideration in the transportation demands serving process. In the second case, the parameters of each route are defined. As Table 3 and Table 4 shows, the parameters of first and second route are defined.

Initial transportation task	Time	Loc$_x$	Loc$_y$	Quantity of task [unit]
1.	8:15	10	8	-20
2.	9:05	13	20	-30
3.	9:33	15	33	48
4.	9:55	25	29	-45
5.	10:20	29	35	-34
6.	10:40	40	50	40
7.	11:16	42	53	22
8.	13:20	48	55	-35
9.	14:40	58	66	-100
10.	15:40	65	80	80

Table 3: 1. Route parameters of the first vehicle

The quantity of each initial transportation tasks, depends on task request (delivery/pick up) are already pre-specified.

Initial transportation task	Time	Loc$_x$	Loc$_y$	Quantity of task [unit]
1.	8:00	2	20	-30
2.	9:30	15	30	28
3.	10:20	23	37	-45
4.	11:10	31	35	-25
5.	12:00	37	42	72
6.	12:20	41	45	-5
7.	13:40	55	51	5
8.	14:10	60	62	-35
9.	14:50	70	74	25
10.	15:20	80	85	-44

Table 4: Route parameters of the second vehicle

The vehicle's capacity impact is analyzed in this case. The first route vehicle capacity is 250 unit and the second is 200 unit.
If the location of the additional task is 11 (42; 48), then we also can find the optimal way. We defined the quantity of new task using RANDBETWEEN function. It returns a random integer between 1 and 50 value. It means the additional demand will affect to which route could be fulfil the request. We consider to the capacity occupancy in routes can be calculated with Eq. (2). The results are depicted in Figure 4.

1. route	DT1 [unit]	CAP1 [unit]	2. route	DT2 [unit]	CAP2 [unit]
1.	51	259	1.	49	199
2.	40	229	2.	32	227
3.	31	277	3.	22	182
4.	25	232	4.	17	157
5.	18	198	5.	8	229
6.	3	238	6.	3	224
7.	5	260	7.	13	229
8.	9	225	8.	23	194
9.	24	125	9.	38	219
10.	39	205	10.	53	175

Figure 4: Calculation results

In case of scenarios, the optimal path can be found depending on the available capacities. Capacity occupancy shows which besides initial tasks would be able to handle a new task.

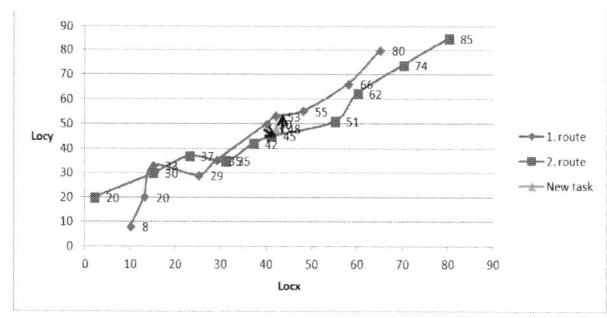

Figure 5: Solution of second case

Figure 5 shows the 6th initial task is the nearest to additional task to fulfil the new additional request.

We can also change the location coordinate of transportation tasks, and then we can assign random value with the mentioned function.

In the case of the scenarios the nth optimal path can be found depending on the available capacities. Capacity occupancy shows which beside initial tasks would be able to handle a new task for any route.

6 Conclusions

The companies can deliver products, and serve customer demands. Routing planning is a difficult task of transportation. Our model can determine the optimal route between transportation tasks, running transportation tasks to additional demands for small number of routes, which take into consideration the transportation operation costs. This tool can be used mainly cases where courier service has fix tasks of customer demands and we can pre-calculate routes. The method only needs some delivery points with locations and quantities parameters of the package. In this mind, the method is available to serve out short-term delivery and pick-up tasks, like GLS, or DPD. It can solve initial and additional tasks, which are connected to supply chain processes. It can determine the capacity occupancy, as design aspect of supply chain and related costs.

In complex supply chain problems should be solved with heuristic and metaheuristic solutions have to be taken into consideration.

This should be also considered in further studies, and design it for a large number of routes.

7 Acknowledgements

This project has received funding from the EFOP-3.6.1-16-00011 "Younger and Renewing University – Innovative Knowledge City – institutional development of the University of Miskolc aiming at intelligent specialization" project implemented in the framework of the Szechenyi 2020 program and the European Union's Horizon 2020 research and innovation programme under grant agreement No 691942. This research was partially carried out in the framework of the Center of Excellence of Mechatronics and Logistics at the University of Miskolc.

8 References

[1] Zhu, R., Zhai, Y.: Research on the application of VRP theory in logistics transportation. MATEC Web of Conferences, 100, art. no. 10005064, 2017.

[2] Ranieri, L., Digiesi, S., Silvestri, B., Roccotelli, M.: A review of last mile logistics innovations in an externalities cost reduction vision. Sustainability (Switzerland), 10 (3), art. no. 782, 2018.

[3] Noyan, N., Kahvecioğlu, G.: Stochastic last mile relief network design with resource reallocation. OR Spectrum, 40 (1), pp. 187-231, 2018.

[4] Perboli, G., Rosano, M.: A decision support system for optimizing the last-mile by mixing traditional and green logistics. Lecture Notes in Business Information Processing, 262, pp. 28-46, 2018.

[5] Chen, Y., Yu, J., Yang, S., Wei, J.: Consumer's intention to use self-service parcel delivery service in online retailing: An empirical study. Internet Research, 28 (2), pp. 500-519, 2018.

[6] Zhou, L., Baldacci, R., Vigo, D., Wang, X.: A Multi-Depot Two-Echelon Vehicle Routing Problem with Delivery Options Arising in the Last Mile Distribution. European Journal of Operational Research, 265 (2), pp. 765-778, 2018.

[7] Kin, B., Ambra, T., Verlinde, S., Macharis, C.: Tackling fragmented last mile deliveries to nanostores by utilizing spare transportation capacity-A simulation study. Sustainability (Switzerland), 10 (3), art. no. 653, 2018.

[8] Deutsch, Y., Golany, B.: A parcel locker network as a solution to the logistics last mile problem. International Journal of Production Research, 56 (1-2), pp. 251-261, 2018.

[9] Ha, Q.M., Deville, Y., Pham, Q.D., Hà, M.H.: On the min-cost Traveling Salesman Problem with Drone. Transportation Research Part C: Emerging Technologies, 86, pp. 597-621, 2018.

[10] Leung, K.H., Choy, K.L., Siu, P.K.Y., Ho, G.T.S., Lam, H.Y., Lee, C.K.M.: A B2C e-commerce intelligent system for re-engineering the e-order fulfilment process. Expert Systems with Applications, 91, pp. 386-401, 2018.

[11] Schermer, D., Moeini, M., Wendt, O.: Algorithms for Solving the Vehicle Routing Problem with Drones. Lecture Notes in Computer Science (including subseries Lecture Notes in Artificial Intelligence and Lecture Notes in Bioinformatics), 10751 LNAI, pp. 352-361, 2018.

[12] Zhang, Y., Shi, L., Chen, J., Li, X.: Analysis of an automated vehicle routing problem in logistics considering path interruption. Journal of Advanced Transportation, 2017, art. no. 1624328, 2017.

DESIGNING THE CONSOLIDATION NETWORK OF USED BATTERIES

Tatiana Solianyk
Department of Information Control Systems/ Aircraft Control Systems Faculty
National Aerospace University "Kharkiv Aviation Institute", Ukraine

Mikhail Dvornikov
Department of Information Control Systems/ Aircraft Control Systems Faculty
National Aerospace University "Kharkiv Aviation Institute", Ukraine

1. Introduction

No one can imagine reality without technical means, devices and mechanisms. Humanity has become extremely dependent on the unique "results" of technological progress, which in most cases need batteries that ensure their performance.

Any battery has its own resource. If it is finished – the battery is not able to perform its functions. Therefore, the batteries must be recycled. If the battery is simply thrown to the dump, the likelihood of harmful components and substances falling into the water, soil and air increases.

With the passage of time, lead, mercury, sulfuric acid begins to flow out of the discarded battery, getting into the soil, and then into the ground water, they cause enormous harm to the environment.

Disposal of storage batteries is a process that is carried out only by competent employees at processing enterprises that have a corresponding license [1].

Along with car batteries, recycling should be subject to batteries that ensure the operation of portable devices, since they contain the same dangerous substances.

Every day our consumers get rid of almost one million batteries, each of which can pollute 20 square meters of soil and 400 liters of water [2]. According to experts, the weight of batteries discharged per day is 12 tons, to better representative - this is a large truck. For the disposal of only 0.1% of these batteries.

With the proper organization of the recycling process, you can get revenues from the sale of lead, recycled plastic and other equally important secondary raw materials.

However, despite the fact that the technological processes of utilization are being improved, and to date, it is possible to almost completely provide waste-free production, many developed countries (for example, Japan, China) are waiting for the moment when the most optimal way of utilizing batteries will be developed. For this reason, they collect batteries in special storage [3].

There are two points of utilization of used batteries on the territory of Ukraine [3]:

- The plant for non-waste processing of used batteries in Dnepropetrovsk. The closed technological cycle practiced at the plant allows processing all components of storage batteries, including electrolyte. The possible volumes of processing of accumulator batteries by the plant is about 3 million copies a year.
- Lviv state enterprise "Argentum", which is engaged in the recycling of batteries and batteries of portable devices. The capacity of this enterprise is calculated at a maximum of 3 tons per month, and the actual volume of processing is 300 kg per month.

Due to the low level of the accumulator batteries collection, these enterprises in Ukraine are almost idle. The reason is simple: the country does not have a system for collecting household batteries and accumulator batteries.

Therefore, the task of collecting and utilizing used batteries is actual and practically meaningful. The given work is devoted to creation of a collection network of the f used batteries and accumulators.

2. Analysis of the utilization process and possible recycling objects

In our time, there are a number of different types of batteries, but the process of their utilization is almost identical, and is provided using similar equipment [3]. The reason for this is that most of the standard batteries consist of similar components: usually a plastic case, internal plates of different alloys (most often active metals) and electrolyte.

The whole process of processing battery batteries involves several successive technological processes: draining and neutralizing the electrolyte, cutting the body of the batteries, separating the plates from the battery case, crushing the batteries with special equipment and then melting secondary raw materials in shaft furnaces.

All the variety of batteries for portable devices can be divided into secondary (rechargeable) and primary (not rechargeable) batteries.

As expected, the increase in demand for secondary batteries (about 85% of the world market) is caused by the growth of the sphere of portable electronics, such as tablets and mobile phones [4].

Demand for non-rechargeable batteries, respectively, falls (about 14% of the world market). They are used in watches, electronic keys, remote consoles, light sensors, signal beacons and in devices for the military industry [4].

The lead-acid system holds its positions, being a reliable and economical power source for the widest applications. The same areas of application of the lead-acid system can be divided into automotive (20%), where starter batteries (also known as SLI) are used, stationary (8%), where batteries are used for backup power, and batteries for driving 5%), for example, for golf carts, strollers or self-propelled lifts [4].

The type of power source will affect the receiving point and its location. For example, for the collection of electric batteries, containers of various sizes can be used, which are conveniently located in public places. The collection points for storage batteries should be located at the car maintenance stations.

All points of reception of used batteries will form a collection network that must meet certain requirements.

3. Requirements for the collection network

Within organizing the collection of used batteries, it is necessary to take into account many different factors, such as the covered territory, the developed transport infrastructure, the material and technical base for organizing the collection of batteries and their transportation, labor resources and so on.

The projected collection network should provide the following functional capabilities:
- Collection of used batteries of a wide range.
- Accessibility for the population. The network should be as close as possible to the potential sources of recycling elements.
- Safety of used batteries. Items should be equipped with special containers that will ensure the physical safety of the items handed over.
- Containers for collection of sufficient size. The size of the container should provide the collection of such a volume of used batteries, which will be expedient from an economic and ecological point of view.
- Necessary conditions and terms of storage. Removal of batteries should be carried out as the container is filled. In the period between exportations, appropriate temperature and humidity parameters should be ensured, which will not lead to adverse consequences.

- Convenience of exporting used batteries from the point. The point of reception should be accessible to the attendants, and equipped with an entrance for the vehicle.
- Delivery to the central storage or processing. The network to be developed must contain transport links that will allow the delivered batteries to be delivered either to the central storage or to processing.
- Effective and economical transportation. The network topology should be constructed in such a way that the organization of routes between receiving points was efficient and cost-effective.
- The cost of arranging the reception point. Like any costs, the cost of organizing the reception point should be economically justified.

4. Algorithm for building a collection network for used batteries

The classical approach of constructing any network structure involves the use of methods of graph theory.

This work uses the capabilities of the simulation system AnyLogistix [5, 6].

AnyLogistix is an easy-to-understand tool, which can be used to address a wide range of supply chain management (SCM) problems. By reducing technical complexity to a minimum, anyLogistix allows to focus on management decision analysis and use KPIs for operational, customer and financial performance measurement and decision-making.

You can model the supply chain in two ways
- Analytical modeling that uses optimization models to investigate the supply chain
- Simulation modeling that uses a set of objects and rules that describe their dynamic behavior and their interaction to represent the supply chain.

The algorithm for building a network contains the following steps.

Step 1. Selection of the territory for collection of used batteries.

The AnyLogistix already contains a GIS map of the world. This map already shows some urban infrastructure, as well as urban transport systems. It is necessary to determine the expected service area.

Step 2. Select the type of the receiving point.

The type of recycling object will affect the source of its occurrence. If we are talking about batteries of portable devices, it is advisable to place them in public places. If we are talking about car batteries - at the service stations.

Step 3. Determine the collection points for used batteries for disposal.

Determine the required number of reception points and set their location.

Step 4. Determine the number of terminals required.

The number of terminals should provide the greatest possible reduction in the through-route routes.

We use the methods of the theory of logistics to calculate the required number of terminals. With end-to-end technology, transportation is carried out between points by one car. On the allocated territory, there are n points of reception of used batteries. Between any of these points, goods can be transported by road. Such transportation of goods between two points is called a possible motor transport connection. The total number of possible motor transport connections depends on the number of points n served by transport:

$$N_a = \frac{n * (n - 1)}{2}.$$

When carrying out the transportation of goods through the terminal system, the total number of motor transport connections is determined by the sum of the number of interterminal connections and connections between terminals and reception points when performing transport operations:

$$N_T = \frac{K * (K - 1)}{2} + \frac{n * (n - K)}{2 * K},$$

$$K = \sqrt[3]{\frac{n^2}{2}},$$

Where K – the optimal number of terminals in the region.

For any region, there is an optimal number of terminals, which provides a minimum number of motor transport links and their maximum freight traffic.

Expected average distance L of delivering goods to terminals:

$$L = 0{,}282 \sqrt{\frac{S}{K * R}},$$

Where S – area of the region, sq. km;
 K – the optimal number of terminals in the region;
 R – The coefficient of development of the road network, equal to the share of the region area, which can be served by road. R varies from zero (for a region where there are no roads) to one (for a region whose entire area is covered by a road network).

Step 5. Optimize the network topology with terminals.

Optimization of topology implies fixing a certain number of reception points for the corresponding terminals, because of which a minimum number of motor transport links between points will be reached and their maximum freight traffic will be reached too. The optimization criterion is the average distance between the collection point and the terminal. Optimization is performed automatically by means of the AnyLogistix.

5. Example of building a collection network for used batteries

As an example, a variant of a network for collecting used batteries of portable devices is shown.

Step 1. The collection network is formed for the city of Kharkov. Its total area is 350 square kilometers. The network being developed should cover 35% of this territory.

Step 2. As the object of recycling selected batteries for portable devices, so it is advisable to place reception points in the chain stores.

Step 3. As the initial points of reception are selected networks of two large supermarkets in the city, whose shops are located in all central and residential areas of the city (Fig. 1). Most of the reception points have already been indicated on the map. Missing points of reception are easy to find at the address and denote.

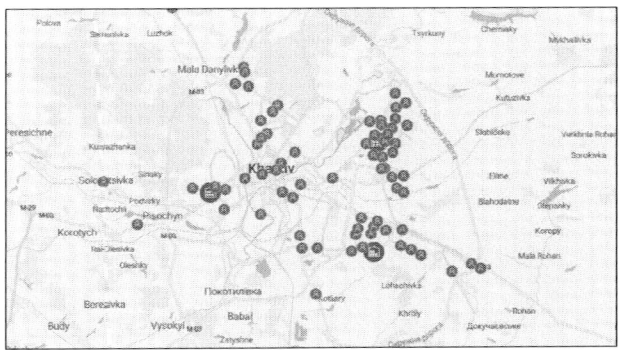

Figure 1: A map of the selected area with preset points for the reception of used batteries

Step 4. Determine the number of terminals required. The initial data and the results of the calculations are given in Table 1.

№	Indicator name	Symbol	Units	Value
Initial data				
1.	Area of the collection	S	Sq. km	298
2.	Number of reception points	n	pc	67
3.	The coefficient of development of the road network	R		1
Results				
4.	Number of possible motor transport links	N_a	pc	2211
5.	Number of motor transport connections with the terminal system	N_T	pc	217
6.	Average distance of cargo delivery to terminals	L	km	5

Table 1: The initial data and the results of the calculations

After receiving the information, set the required number of terminals in the system. (Figure 2.)

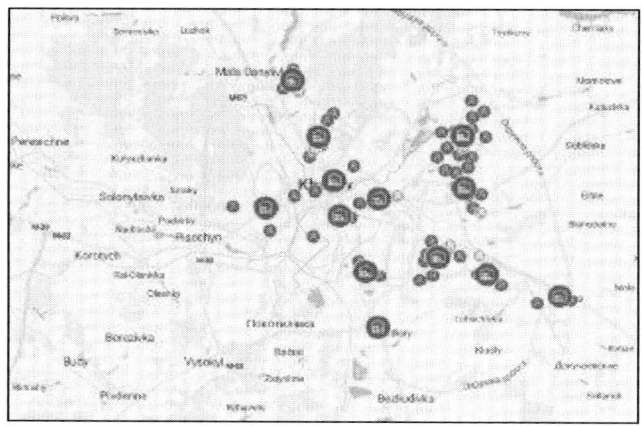

Figure 2: Map of the selected area with preset points for the reception of used batteries and terminals

Step 5. We form the network - we fix the possible number of reception points for a certain terminal. As a criterion for optimizing the network topology, we use the average distance between the collection point and the terminal. The results of building the network are shown in Fig. 3.

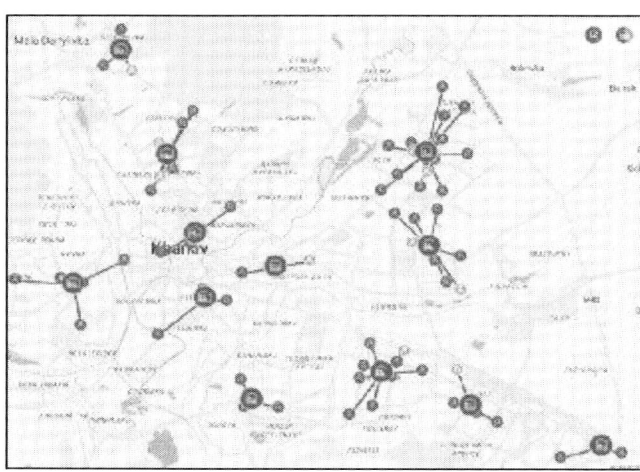

Figure 3: optimized collection network for used batteries

The resulting collection network is one of the possible options for its construction. In the original model, it is easy to change the number of collection points. Initially, the system automatically arranges terminals. If necessary, their location can be adjusted manually, taking into account additional information (for example, the exact location of the existing office).

Similarly, you can build a collection network of car batteries. As reception points, it is advisable to use technical service stations.

6. Conclusions

In this paper, the task of designing a network for collecting used batteries is solved. A distinctive feature is the use of AnyLogistix, which allows you to build a model with a large number of elements, as well as provides automation of labor-intensive design stages. In the future, the obtained results can be used as an independent model, as well as for integration with another application (for example, AnyLogis) in the integrated simulation of the process of organizing the disposal of used batteries.

7. References

[1] «Tekhnologiya utilizatsii starykh akkumulyatorov», *http://avtowithyou.ru/remont-avtomobilej/texnologiya-utilizacii-staryx-akkumulyatorov/,* December 23, 2016.

[2] "Sbor i utilizatsiya batareyek v Ukraine", *http://blagoustriy.info/experiences/15/show/,* February 14, 2014.

[3] Sergey Vol'ter "Pravil'naya pererabotka akkumulyatornykh batarey", *http://www.electra.com.ua/akkumulyator/299-pravilnaya-pererabotka-akkumulyatornykh-batarej.html,* October 10, 2013.

[4]   Abramova Olesya "Analiz mirovogo rynka elektricheskikh batarey", *https://best-energy.com.ua/support/battery/695-bu-103*, February 29, 2016.

[5]   Ivanov D. (2017) Supply Chain Simulation and Optimization with anyLogistix. HWR Berlin, available at https://www.anylogistix.com/upload/alx-book.pdf

[6]   https://www.anylogic.com/resources/books

DEVELOPMENT OF A UNIQUE CATALOGUE OF CRITERIA TO COMPARE INTERNAL MATERIAL AND URBAN TRANSPORT SYSTEMS

M.Sc. Franziska Thomas
Institute of Logistics and Material Handling Systems
Otto von Guericke University Magdeburg, Germany

Univ.-Prof. Dr.-Ing. Hartmut Zadek
Institute of Logistics and Material Handling Systems
Otto von Guericke University Magdeburg, Germany

1 Introduction

1.1 Motivation

The consequences of increasing urban transport as air pollution and noise emission get a rising impact on daily life and the health of citizens. The biggest part of these impacts are produced by transport of goods. Therefore many projects had focused on the minimization or shift of goods transport in the 1990ies. Only a small number of the former concepts generated by these projects are still in action. Because of growing traffic there are more researches required.

For generating new possible solutions it's a proven remedy to search for related systems and problems and adapt these solutions. So similar systems to the urban transport ones have to be found. If the traffic flow is considered as material flow and cities as companies, internal material flow systems can be regarded as references. But to transfer concepts the analyzed systems and problems have to be compareable.

This paper focuses on the development of homogeneous criteria which enable a system syncrisis. Beside of the uniform system characteristics requirements on logistical systems are involved to generate a unitary system description. Hence, urban transport and material flow systems are able to match.

1.2 Already transferred concepts

Different concepts which have already been transferred between both systems seem to demonstrate that a transfer is possible. For example trams in cities were used for the transportation of goods. The concept of supply through a regular transport route and standardized trailers can be found in Intra logistics as the tugger trains. It usually consists of a traction unit and several transport racks, which can individually (un-) loaded. This is already used to allow a production without fork-lift trucks and lean processes in the material supply. This concept is already employed as the CargoHopper in the Netherlands. The first variation, named as 'CargoHopper I', consists, as the tugger train, of a traction unit with several trailers, which has the task to supply the protected areas of inner cities [1]. This suggests also other (supply) concepts can be transferred. To scrutinize this, both systems have to be compared to identify possible transferable concepts. For this a unitary system description has to be generated.

1.3 Transferability between material flow systems

Even in literature hints can be found, that transportation concepts are able to transfer. According to Krampe, solutions for logistical problems in regional areas are possible to generate both from methods for traffic and material flow planning [2, p. 433]. Also Arnold agrees that problems in internal and external material flow systems can be solved by similar methods [3, p. 6]. Correspondent to Clausen the focus in traffic planning is set on the material flow [4, p. 4]. In conjunction with the previous statements this suggests that it is possible to transfer these concepts.

2 Development of systemic criteria

In the following it has to scrutinize how both systems can be compared and therefore identify transferable concepts. Hence the objective is to uniformly describe both systems. Based on this description similar characteristics are identified and therefore concepts can be transferred. During this, a systemic comparison according to Stykow to identify unitary criteria is performed. For these criteria, several specifications are determined to classify systems [5, p. 38].

2.1 System characteristics

According to Brandau systems in general consist of the following characteristics:
- Systems intention
- Elements
- Relations
- Systems environment
- Systems interfaces
- System input and output variables [6, p.17-18]

The intention defines the context and the way a system is regarded. Elements are the components of a system, connected by relations. The system components and the system environment are marked out by the system boundary. System interfaces enable the exchange of input and output variables between the elements and the systems environement.

the characteristics have to relate to logistical systems.

2.3.1 Systems intention

Logistical systems are service systems, hence they focus the processes. Therefore the processes in logistical systems have to be identified. They are deduced from the basic material flow functions *producing, moving* and *resting* mentioned by Martin [7, p. 22].
Ten Hompel defines material flow processes as processes, where material objects are transferred regarding time, place, quantity, composition or quality [8, p. 7]. Table 2 illustrates the assignment of these specifications to the material flow functions. It demonstrates that the processes *checking, storing* and *unintentional storage* have the same impact on the specifications. Therefore they can be combined to the process 'storing'. The resulting material flow functions are equal to

System criteria	Examples for the system 'urban logistics'	Examples for the system 'internal material flow'
Systems boundary	– City boundary – District boundary	– Company boundary
Elements	– household – agencies – companies – commerce – persons – goods – means of transport	– departments – production halls – areas – work stations – goods – means of transport
Relations	– streets – railway – waterway – air transport	– streets – railway
Systems intention	– Transport of goods – Enable passenger mobility	– Transport of goods
Systems interfaces	– Streets, city boundaries	– Goods receipt – Input and output gates
In- and output variables	– Goods incl. means of transport – Passengers (incl./ excl. means of transport) – Information	– Goods incl. means of transport – Information

Table 1: examples for the system criteria for the regarded systems

2.2 Apply the characteristics to the systems

Table 1 contains examples of both regarded systems. It's used for illustration only and not exhaustive. Several characteristics have equal specifications (i.e. systems intention, connections, in- & output variables). But there are also some differences. Further classifications of these characteristics have to be done to scrutinize if these differences also exist after closer inspection and therefore prevent the transfer of concepts.

2.3 Determination of the specifics of the characteristics

Preferring to systems intention and classification by Brandau the regarded systems can be assigned to logistical systems [6, p. 19]. Hence

the functions mentioned by Arnold [3, p. 8].

2.3.2 System boundary

The system boundary of logistical systems is either a physical or a logical one. For example the plant borders or the walls of a factory hall are physical boundaries. But smaller subsystems often aren't able to get separated by physical or 'real' boundaries and a logical border is needed. A further property of the system boundary is its permeability. It can be an impermeable, semi-permeable or a permeable border. Whereas the permeable border allows every unit to cross this border in every direction, the impermeable border allows none of them. The semipermeable border has some restrictions, as the kind of units which can pass the border or the direction.

Process	Location	Time	Quantity	Composition	Quality
Operate	(X)	X	(X)	X	X
Check		X			
Transport	X	X			
Handle	X	X	X		
Store		X			
Unintenional stay		X			

Table 2: assignment of the material functions to the properties of goods in systems

2.3.3 Elements

The elements in logistical systems are named logistical objects. According to Jünemann they can be separated into two groups. The first one consists of those, who can be transformed and transported through the system. The second one

2.3.4 Relations
The relations are determined by the different modes of transport:
- streets
- railswas
- waterways
- air transport
- pipes

Systems criteria	Specifications						
Border							
- physicality	Logical				Physical		
- permeability for elements	(material) goods		Persons		Energy		Information
-direction of permeability	Inwards				Outwards		
Processes	Transportation		Handling		Storing		Operating
Mobile elements	(material) goods		persons		Energy		Information
Immobile elements	means of production		means of transport		means of handling		means of storage	control units
Relations	Streets	Railways	Waterways	Air transport	Pipelines	Wire	Radio network

Table 3: developed systemic criteria to compare systems

consists of elements/ objects which perform these transformations [9, p. 3]. Lucke differentiates between mobile and immobile elements [10, p. 38]. The mobile elements are assigned to those elements, which are transformed or transported through the system. Otherwise the immobile elements can be designated as the performing elements. This differentiation can be used to identify the relevant mobile elements from the logistical objects. Therefore the following logistical objects are applied to the mobile elements:
- (material) goods
- persons
- energy
- information

The immobile elements are performing elements, therefore they perform the in chapter XX mentioned processes.

In conjunction with the transformed mobile elements, the corresponding means of material flow can be identified:
- means of production
- means of transport
- means of handling
- means of storage
- control units

Kummer also mentions communication traffic as a mode of transport [11, p. 40]. To stay at an infrastructural point of view, the wireless and the tethered communication are added to the relations. All of the mentioned characteristics and specifications are combined in a morphological box. It is visualized in Table 3.

3 Determination of the criteria due to the requirements

To compare both systems, the systemic criteria mentioned below don't suffice. Therefore the systems requirements for logistical systems have to be identified. Rupp defines requirements as properties that are necessary for persons or systems to reach specific objectives. Possible objectives are either solving a problem or reaching several specifications [11, p. 13]. According to the systems treatment on logistical services, the following parameters can be eliminated from this definition:
1. Requirement guideline (objectives)
2. Submitter of the requirements (Who specified the objectives?)
3. Subject of requirements (Whereby are

requirements presented?)

4. Requirements (Which properties are demanded?)
5. Wo has to fulfill the requirements?

3.1 Requirements on logistical systems

The aim of logistical systems is to provide logistical services, which can be described by the following goals:
- Right goods
- Right quality
- Right quantity
- Right date
- Right location
- Right costs
- Right information
- Ecological correctness

These goals of the logistics are representing the requirement guidelines. They are mostly submitted by the customer. Only the last two requirements are mostly submitted by further involved parties. For the information, additional involved parties are the shipper and the carrier. The ecological correctness is mainly claimed by the parties in the systems environment, which are directly concerned by the transport. The subject of requirements can be classified in a similar way. Whereas the first two goals are obtained on the goods, the quantity, the date and the location focus the supply of goods. The costs and ecological impacts are specifications of transport services. They evaluating the supply from another perspective, therefore they are regarded in a special manner. On a completely different way the information have to be seen. Whereas the pervious requirements are focused on the material flow, the information focus on the simultaneously operating flow of information.

3.2 Systemic treatment of the requirements

After applying the requirements to the involved parties, these parties have to be assigned to the systemic characteristics.

The major part of these requirements can be assigned to the processes. Only the first two requirements are submitted to the mobile elements. The immobile elements have to fulfill all requirements. Most of the requirements are set by the customer. They can either be part of the system or part of the systems environment, belonging to the focus of the regarded system. In the former case the customer has to be assigned to the immobile elements. The ecological evaluation is especially done by the systems environment, therefore not by directly involved elements.

Since the requirements on the systems have to be identified, the mobile elements and the transport and supply process have to be studied. Through the first two requirements are specific for transport and therefore have a higher impact on the concepts, they will be excluded for further considerations belonging systems comparison. Thus the supply with the components quantity, date and location is used for further comparisons.

3.3 Requirements on material supply

Bullinger developed a classification to select specific strategies for the material supply, which is taken to represent the requirements on the systems [13, p. 17]. The type and the quantity of supply can be assigned to the requirement on the quantity. The source and drain of supply can be attached to the requirements on the location. The requirements on time are considered by Bullinger indirectly through the mode of supply. The activation and execution can't be assigned to any requirement mentioned below. But they enlarge the comparison by the submitter and fulfiller of the requirements. The wording used by Bullinger is focused on internal material flow systems. Therefore the 'work station' is replaced by 'point of use'. The summary of the requirements is illustrated in Table 4.

Criteria	Specifications				
Mode of supply	According to command			According to consumption	
Type of supply	Collected orders	Entire order (batch size> 1)	Part of an order	Single product (batch size = 1)	Part/ component
Supplied quantity	Bundle oriented			Exact number of units	
Source of the supply	Striking distance to the point of use	In the system of point of use (interim storage)		In the system of point of use (neutral warehouse)	Other upstreamed areas
Drain of supply	To point of use		Near to point of use		In the system of point of use
Activation of the order	Bring-principle				Take principle
	Superordinate, central system	Decentral, upstream		Decentral, adviser of point of use	Decentral, point of use
Execution of the order	Bring-principle				Take principle
	Supplier of the systems		Upstream activities		System of the point of use

Table 4: criteria to compare systems taken off the material supply [according to 13, p. 17]

4   Conclusion

It's been the obstacle of this paper to identify unique criteria to compare both systems 'intra logistics' and 'urban logistics'. At first the systemic description was used to determine general system characteristics. With the aid of specialist literature the specifics for the system characteristics for logistical systems had been identified.
Thereafter the requirements had been classified by using the goals of logistics. These requirements were assigned to the system characteristics to highlight the relevant requirements. Based on this analysis the proof was given, that the classification for material supply by Bullinger can be used to enlarge the comparison.

5   References

[1]   Randelhoff, Martin. Zukunft Mobilität. Cargohopper: Das Fahrzeug für eine stadtverträgliche, flächeneffiziente und schadstofffreie Innenstadtlogistik. https://www.zukunft-mobilitaet.net/120226/konzepte/innenstadtlogistik-cargohopper-konzept-staedtischer-lieferverkehr-elektromobilitaet-ohne-stau/

[2]   Krampe, Horst. (2012): Wirtschaftsverkehr in Ballungsräumen. In: Krampe, Horst (editor): Grundlagen der Logistik. Einführung in Theorie und Praxis logistischer Systeme. München: Huss-Verlga, p.413-448.

[3]   Arnold, Dieter; Furmans, Kai. (2007): Materialfluss in Logistiksystemen. Berlin, Heidelberg: Springer-Verlag.

[4]   Clausen, Uwe. (2013): Einführung und Begriffe. In: Clausen, Uwe; Geiger, Christiane (editors): Verkehrs- und Transportlogistik. Berlin, Heidelberg: Springer Vieweg, p. 3-5.

[5]   Stykow, Petra. (2007): Vergleich politischer Systeme. Paderborn: Wilhelm Fink.

[6]   Brandau, Annegret; Schenk, Michael. (2015): Ganzheitliches Konzept zur Modellierung und Analyse von Zustandsdaten logistischer Objekte. Magdeburg: OVGU Magdeburg.

[7]   Martin, Heinrich. (2011): Transport- und Lagerlogistik. Planung, Struktur, Steuerung und Kosten von Systemen der Intralogistik. Wiesbaden: Vieweg + Teubner.

[8]   Ten Hompel, Michael; Sadowsky, Volker; Beck, Maria. (2011): Kommissionierung. Planung Und Berechnung Der Kommissionierung in Der Logistik. Dordrecht: Springer.

[9]   Jünemann, Reinhardt; Schmidt, Thorsten. (1999): Materialflußsysteme. Systemtechnische Grundlagen; mit 36 Tabellen. Berlin: Springer.

[10]   Lucke, Hans-Joachim. (2012): Systemtheoretische Grundlagen der Logistik. In: Krampe, Horst (editor). Grundlagen der Logistik. Einführung in Theorie und Praxis logistischer Systeme. München: Huss-Verlag, p. 37-56.

[11]   Kummer, Sebastian. (2010): Einführung in die Verkehrswirtschaft. Stuttgart: UTB GmbH.

[12]   Rupp, Chris. (2013): Systemanalyse kompakt. Berlin, Heidelberg: Springer.

[13]   Bullinger, Hans-Jörg; Lung, Martin M. (1994): Planung der Materialbereitstellung in der Montage. Stuttgart: Teubner.

GENETIC ALGORITHM IN COMPARISON TO ABC ANALYSIS FOR WAREHOUSE PICKING AREA LAYOUT CALCULATION

Aleksandrs Avdeikins
Transport and Telecommunication Institute, Riga, Latvia

Andrejs Simakovs
Trialto Latvia Ltd.
Saulgozi, "Dominante", Latvia

Mihails Savrasovs
Transport and Telecommunication Institute, Riga, Latvia

Warehouse business processes have several KPI's – one of them is picking efficiency. If to omit human factor and the technical equipment of the warehouses, picking efficiency is most affected by the correct goods placement – the better the goods are located, the shorter will be the picking distance for each order. It means that individual orders will be picked faster. Usually to determine the correct location for the goods 3PL's are using ABC analysis, that includes indicators like count of orders, goods turnover, picking rate, weight etc. There are also more complicated indicators like goods adjacency. Such indicators are hard to take into account using ABC analysis, as it requires sophisticated analysis of customer orders.

The proposed approach is to use a genetic algorithm to find the better placement of the goods as it does not require to use indicators that are hard to acquire. We believe, that this approach could minimize a human mistake ranking the selected indicators as it is required in ABC analysis.

The goal of this paper is to compare calculated picking area layout using both methods and determine which approach is better in terms of picking efficiency and each method practical usability.

Keywords: 3PL, Logistics, ABC analysis, Genetic Algorithm

1 Introduction

The growth of third-party logistics (3PL) companies has generated the need for an accurate costing system, to prevent distortion of the information produced by traditional costing systems [1]. Furthermore, the competition between these companies generated the need for more accurate financial and non-financial information, for them to gain an insight in logistics activities, and either eliminate unprofitable activities or improve or re-engineer others [3], which in turn results in lower costs and more competitive prices [2].

The research objective is to present a specific solution to find the better placement of the goods, evaluate current warehouse situation and to improve picking strategy. The solution implies the use of evolutionary algorithms, for determining the optimal location for the SKU (stock keeping unit). The use of an improved picking strategy leads to lower picking costs and maximizing of the picking efficiency.

The research question is whether the genetic algorithm can be used to find an optimal goods placement aiming to decrease picking and human resources costs. The research methodology is based on statistical data collection for client orders on the 3PL warehouse, their analysis and presentation of findings on recommended solutions to increase picking efficiency and overall supply chain performance.

The objective of the research paper is to apply the genetic algorithm for placing SKU in the warehouse, compare it to the legacy ABC analysis method and indicate both methods pros and cons. The designation of the location of the SKU into the warehouse will determinate the travelled distances for the routes made to ensure that the starting and the ending point of each order is dispatch area. Similarly, when positioning the heavier products, or the most fragile, nearest places on the expedition seeks to reduce the effort of moving, as, premature wear of handling equipment. So, when allocating products, that have the flow of movement/demand into the warehouse in places near the areas of the expedition, reduces the total movement [4]. An important operational decision in warehouses is to determine the best storage location for each product to minimize the total material handling effort (or cost). This problem is known in the literature as the storage location assignment problem (SLAP). The objective of SLAP is to determinate in which area/shelf store items of received products, so that the total operational costs are minimized, reducing the movement of materials and according to Carlo and Giraldo [5], the SLAP can be classified according to the information provided about the arrival and departure of the stored products into the warehouses, based on:

Information of the item (SLAP /II): in this case, the information about arrival and departure of the item for the warehouse, are provided, not considering the specifics characteristics about the item, as code, or type of the item;

Information about the product (SLAP /PI): for this classification, it is known product data, as code, batch, unit value, among other items. Thus, held the attribution of classes for the product. The products will be storage, according, with pre-established areas for a given class;

No data (SLAP /NI): do not exist a relationship of storage with item/product, since no information about the classification above is considered. In this case, the storage occurs randomly obeying the closest position. The information and necessary criteria for the application of SLAP for the definition of physical arrangement are related to several factors as the floor plan of the warehouse, area/capacity of storage, the relationship of total positions of shelving, as data about the products to be stocked and order processing time.

2   Case study

The paper covers simplified model analyzing only picking places as most of the order lines picked from there. Trialto Latvia Ltd., the big 3PL operator in Latvia, provided client order placement information and warehouse layout for the analysis. Figure 1 shows current warehouse layout where research results will be applied.

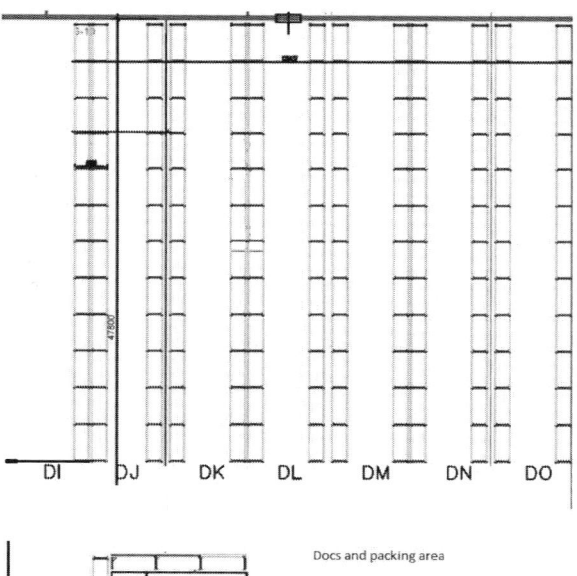

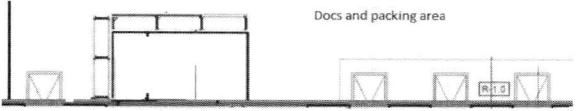

Figure 1: Logistics warehouse layout

Currently for the items placing inside the warehouse Trialto Latvia using ABC analysis. This paper research will try to find alternate better way to find the best location for the item using GA (genetic algorithm) to minimize picking time and costs. The investigated warehouse used for storage of the household goods. The time period from which source data originate encompassed the time from February 2017 to April 2017. The warehouse comprises 974 inventory items for ABC analysis and GA calculations.

3   ABC method approach

The ABC method is usually employed to investment management in stock, but, also can be used for management of operational activities. The ABC Method has been largely applied in the literature. It enables to reduce the total distances of goods movement in the warehouse. The technique shows a great efficiency and has some similarities with the Pareto's Principle. Once the items have been prioritized, a modified Pareto's Principle (commonly referred to as the "80/20 rule") is used to stratify the parts into categories. Typically, three categories are used: "A", "B", and "C", hence the name "ABC Method" [6].

Products are divided into three classes (A, B, C) according to the index of movement. The items with a great index are classified in to class A and allocated in positions near the entry door and/or exit of merchandise; those that have an intermediate index of movement are classified in to class B and allocated in positions with distances; finally, the items classified in class C are those that have the smaller necessity of movement. Therefore, class C is allocated in the positions with higher distances to the port of entry and output of the merchandise [7].

The analysis focused only on the picking locations. The goal of the investigations was to find the best location to place inventory item to minimize picking distance and time. The obtained results are supposed to use for planning picking places of the items. To realize ABC analysis, following order criteria, were determined. These criteria include:

- average count of picks from location,
- weight of the item,
- item volume

Picking figures represents the average count of picks of item per day.

According to the above-mentioned, table 1 shows criteria for considerations of ABC analysis to determine item group and item location inside its group. As main criteria for ABC was used a quantity of picks. To sort items inside the group secondary criteria were used – weight and volume.

Item	Group	Average count of picks	Volume	Weight
Item 1		X1	V1	W1
Item 2		X2	V2	W2
....		...	...	....
Item N		XN	VN	WN

Table 1: Criteria selected for ABC analysis

Examination of physical warehouse configuration gave the possibility to split locations by groups A, B and C. Actual split is presented in table 2.

Group	Locations count
A	126
B	158
C	690

Table 2: Warehouse locations split into groups

"A" group locations defined for the items with the biggest turnover. Lightweight items from the group placed closer to the dispatch area, not to be damaged by heavier SKUs. Group "B" located in the middle area with bigger distance from dispatch area but still can be accessed fast by pickers. Group "C" is the biggest group for slow turnover items. Pickers should take more time to pick items from the location.

3.1 ABC method results

ABC analysis was performed on real sales order data with 8802 orders including 117447 order lines and 974 unique items. SQL server was used to store and analyze data. Table 3 shows ABC analysis output which will be used to calculate its fitness function to compare it to GA results.

Location	Group	Number of picks	Weight	Volume
1	A	310	730.8	34.56
2	A	340	641.34	40.32
3	A	582	641.34	40.32
4	A	382	352.8	0.0576
5	A	370	349.2	0.864
6	A	503	348.48	276.48
7	A	482	348.48	276.48
8	A	538	336.96	276.48
9	A	547	336.96	276.48
10	A	1323	336.96	276.48
11	A	1075	336.96	276.48
12	A	578	336.96	276.48
13	A	861	336.96	276.48
14	A	976	336.96	276.48
15	A	464	336.96	276.48
16	A	392	336.96	276.48

Table 3: Example of results of ABC analysis

4 GA approach. Travelling salesman problem

Evolutionary algorithms are the ones that follow the Darwin concept of "Survival of the fittest" mainly used for optimization problems for more than four decades [8].

Genetic algorithm (GA) is a search and optimization technique that mimics natural evolution. GA has already a relatively old history since the first work of John Holland on the adaptive systems goes back to 1962 [9].

Genetic algorithms are by nature adaptive optimization algorithms that mimic the process of natural selection and genetics [10].

In GA terminology, a solution x is called an individual or a chromosome. Chromosomes are made of discrete units called genes. The main operations of a GA are: selection, crossover and mutation.

The Travelling Salesman Problem or the TSP is a representative of a large class of problems known as combinatorial optimization problems [11]. The problem with warehouse picking is very similar to TSP, as every picker picking goods for each order is like a salesman that needs to find the best route.

The most popular practical application of TSP are: regular distribution of goods or resources, finding of the shortest of customer servicing route, planning bus lines etc., but also in the areas that have nothing to do with travel routes [12].

4.1 Chromosome Representation

Each chromosome coded to represent a solution for the defined problem – warehouse layout. Each gene is a unique item, that position in the chromosome represents the picking order of the items in the warehouse. If there are n items in the warehouse, so the chromosomes will look like shown in table 4.

Chromosome 1:				
Item1	Item2	...	Item(n-1)	Item(n)
Chromosome 2:				
Item22	Item105	...	Item(n-1)	Item(n)

Table 4: Item representation in chromosomes

Each chromosome gene order is the warehouse picking sequence, a route that will be used to pick each order.

4.2 Initial Population, fitness and selection

The initial population consist of 100 randomly generated individuals (chromosomes). Each chromosome is randomly filled with n items and each gene is unique (permutation encoding), as there is no need to have duplicate picking locations for items.

The fitness of each chromosome is calculated as the sum of maximal picking distance for each order. Each order picking distance is calculated from 0 to the furthest item picking location, where distance is item's position in the chromosome.

Distance units of measure is an integer value that for the first item is 1, for second is 2 and etc., increasing by 1 from one SKU to another. Every chromosome fitness function calculated with equation 1.

$$f(C_x) = \sum O_k d_{max}(I) \qquad (1)$$

Where O is order from the set {1,2...k}, and the $d_{max}(I)$ is the distance to furthest picking position. Selection operation is performed based on the fitness function of each chromosome. This operation is required so that strongest individuals would participate in the crossover to get the better children (next generation). As it is required to minimize f, it was selected only 30% of the population, that have f -> min for crossover. Besides of granting a crossover change only for best current population chromosomes, the probability of participating in crossover is higher for the best of the best.

4.3 Crossover and Mutation

To create the next generation of the population an Order Crossover (OX) is chosen as it is used for chromosomes with permutation encoding [13]. The process starts by choosing two crossover points. It copies the subsequence of permutation elements between the crossover points from the cut string directly to the offspring, placing them in the same absolute position [14]. To fill the rest of chromosome a sliding motion is applied. Crossover principal on 10 genes example presented in figure 2.

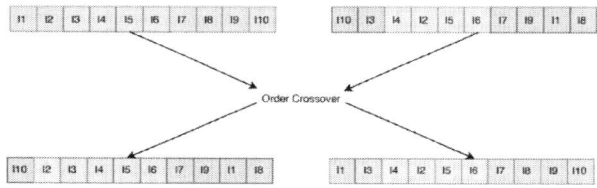

Figure 2: Crossover example

Selection and crossover are evolutionary operations that improve next population step by step, but there is a third operation that can drastically change chromosome's fitness. It is called mutation. It should be used precoitally as this operation can change chromosome fitness to good or to worse, nevertheless, it adds a new solution to population, to help the algorithm not to stuck in local optimum. It tends to mutate 5% of each new population.

4.4 GA simulation results

As input parameters of GA were used 974 unique items from 8802 orders with 117447 order lines. The initial population were 100 chromosomes. After 100 iterations, in case of average fitness function minimization result was less than 1% for last 10 iterations, the algorithm stops. Table 5 shows a summary of input parameters for GA algorithm.

Input parameter	Input value
Population	100
Chromosome length	974 genes
Fitness function	8802 orders
Stop rule	Fitness improvement less than 1%
Simulation max iterations	1000
Simulations count	10

Table 5: GA input parameters

Each of the simulation finished on 1000 iteration and the best chromosome fitness score indicated in table 6.

Simulation Nr.	Number of iterations	Best fitness function result	Run Time
1	986	483764	03:30:00
2	962	416557	03:30:00
3	964	508526	03:55:00
4	957	438915	02:15:00
5	998	476314	02:13:00
6	956	523760	02:16:00
7	827	525924	06:05:00
8	973	541549	06:10:00
9	981	512207	06:05:00
10	984	504699	06:15:00

Table 6: GA simulation running results

The best fitness score GA achieved in the second simulation with 962 iterations (generation). Results for the best simulation shown in figure 3, where X is iteration number, Y is fitness score and Z is chromosome index.

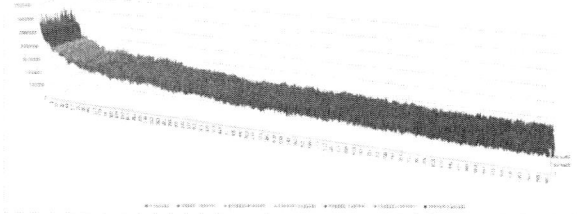

Figure 3: Best GA simulation progress

The graph on figure 3 illustrates that the solution was constantly improved. GA simulation fitness improvement represents an exponential decay function like experience curve (sometimes called Henderson's Law). All 10 runs have similar fitness improvement curve, but with different learning rate.

Figure 4 demonstrates the 2D algorithm progression, where X is iteration number and Y is fitness score.

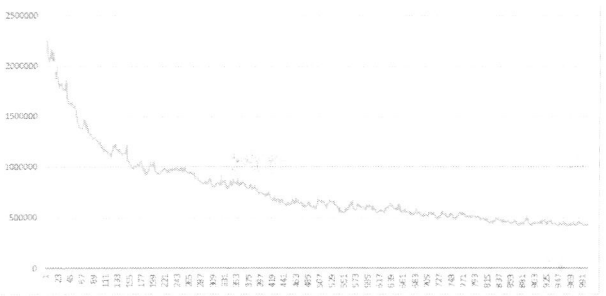

Figure 4: GA evolution

The best fitness function score 416557 calculated for chromosome at 962 iteration. In the real world, that means the optimal items locations for the given order structure were found.

5   GA in comparison to ABC analysis

To proof the practical use of GA for placing items on the warehouse, a comparison was performed between actual warehouse layout, results of ABC analysis and best simulation of GA. To make it comparable to the actual warehouse layout and ABC analysis was calculated fitness function which means total picking time for all orders in the given period. Table 7 shows comparison results. Picking costs were calculated as ratio of ABC analysis and GA results towards to the actual items layout

Layout	Fitness score	Picking costs
Actual layout	2198540	
ABC analysis	1422933	-35%
GA best simulation	416557	-81%

Table 7: ABC analysis and GA comparison versus actual layout

Placing items by ABC analysis will give 35% improvement in total picking time. In case of GA simulation result improvement could be up to 80 %.

6   Conclusions

The purpose of this research paper was to find an alternate and more efficient way for items placement in the warehouse. During research was developed, tested on real data and compared versus ABC analysis GA for the optimal placing of items.
GA simulations were run parallel on different workstations. Each GA simulation took different time to run. In total for processing of 10 simulations 42 hours of computing time spent, which indicates how costly this process is. Due to process cost and warehouse layout change costs, it is advised to use a proposed approach not more than once a month.
ABC analysis and GA have its own pros and cons. Summary of advantages and disadvantages of using above-mentioned methods presented in table 8.

ABC analysis	
Pros	Cons
Simple and fast to develop	Not the best result
Representable	Gives only single solution
Average result	Requires a decision maker
Genetic Algorithm	
Pros	Cons
A better result in comparison to ABC	More complex development and maintenance compared to ABC. Need to hire IT professionals
Gives more than 1 solution	Requires a lot of computing power and several runs
Does not require a decision maker	

Table 8: Pros and cons of ABC analysis and GA

Still, there is a big area for GA improvement and development. For further research, it is planned to perform clusterization and classification for items before running ABC analysis. New parameters will be included in the fitness function. In the future warehouse configuration change costs will be taken into consideration. Adding such will help to avoid situations when GA propose to change items on the locations which are too close and there is no any practical advantage out of this change.
Already at this stage research results of this paper allowed to implement developed GA in Trialto Latvia Ltd. WMS system for monthly analysis of orders flows and item placement.

7   Acknowledgements

This work has been supported by the ALLIANCE project (http://alliance-project.eu/) and has been funded within the European Commission's H2020 Programme under contract number 692426. This paper expresses the opinions of the authors and not necessarily those of the European Commission. The European Commission is not liable for any use that may be made of the information contained in this paper.

8   References

[1]   Griful-Miquela, C. (2001): Activity-Based Costing Methodology for Third-Party Logistics Companies. International Advances in Economic Research 7: 133-146.

[2]   Gunasekaran, A.; Marri, H.B.; Grieve, R.J. (1999): Justification and implementation of activity based costing in small and medium-sized enterprisesî. Logistics Information Management 12: 386-394.

[3]   Pohlen, T.; La Londe, B.J. (1994): ÊÏmplementing activity-based costing (ABC) in logisticsíí. Journal of Business Logistics15: 1-23.

[4]   Tinelli, L.M.; Vivaldini, K.C. T.; Becker, M. (2014): Product positioning optimization in intelligent warehouse. ABCM Symposium Series in Mechatronics 6: 633-643.

[5]   Carlo, H.J.; Giraldo, G. E. (2010): Optimizing the rearrangement process in a dedicated warehouse. In: ELLIS, K.P. et al. Progress in material handling research. Material Handling Industry of America: Charlotte, pp. 39-48.

[6]   Tinelli, L.M.; Vivaldini, K.C. T. (2012): Product positioning optimization using ABC method. ABCM Symposium Series in Mechatronics. 5: 842-848.

[7]   Askin, R.G.; Standridge, C.R. (1993): Modeling and analysis of manufacturing systems. New York: John Wiley and Sons.

[8]   Holland, J. (1975): Adaptation in natural and artificial systems. Ann Arbor: University of Michigan Press.

[9]   Holland, J. (1962): Outline for a logical theory of adaptive systems. J. Assoc. Comput. Mach. 9(3), 297–314.

[10] Goldberg, D.E. (1989): Genetic algorithms in search, optimisation, and machine learning, Addison Wesley Longman.

[11] Greco, F. (2008): Travelling Salesman Problem. Croatia: In-Teh.

[12] Brezina, I.; Cickova, Z. (2011): Solving the Travelling Salesman Problem using the Ant colony Optimization. Management Information Systems 6, No. (4).

[13] Davis, L. (1991): Handbook of Genetic Algorithms. New York: Van Nostrand Reinhold.

[14] Sivanandam, S.N.; Deepa, S. N. (2007): Introduction to Genetic Algorithms. Springer.

IMPACT OF INDUSTRY 4.0 ON AUTOMOTIVE SUPPLIER SYSTEMS

Gábor Nagy, PhD student
Institute of Logistics
University of Miskolc, Hungary

Ágota Tóth Bányainé, PhD
Institute of Logistics
University of Miskolc, Hungary

Béla Illés, Prof. Dr.
Institute of Logistics
University of Miskolc, Hungary

1   Introduction

In the automotive industry, through the strengthening of international competition, the primary objective of the production companies is to meet customer demands at a higher quality and at a lower price. Companies that recognizes the mutual dependency relationship between their customers and their suppliers and can work together with their suppliers and customers to meet customer needs properly may be victorious in the increasingly intense market competition.

Tracking a cross-company relationship system is a very difficult task. Appropriate and quick response to any mistakes on the market is a great advantage as costs can be better planned. The collection and processing of good quality and quantity of data requires a high level of IT and automation. Industry 4.0 is an effective solution to increase the efficiency of manufacturing systems and supply chains.

2   The importance of stocks in the supply chain

For automotive companies operating in a highly complex and uncertain environment, one of the important conditions for their business success is to organize their supply chains to make them work the most efficiently. Today we are talking about global companies and supply chains, as they cover a number of continents, but large geographical distances make the supply chains vulnerable.

The design of supply chains in relation to the operation of production and service processes is becoming increasingly important as logistics costs represent a significant part of the price of goods and services [1]. The development of the stock status and the growth of its economic role justify the need to organize supply chains. Inventories may develop within moments because of rapid technical development or shortening the product life cycle. Stocks, on the one hand, increase sales and production lead times, capitalize on significant capital, provide a great deal of effort to manage and store them, and on the other hand, sub-processes to provide a convenient link and thus help the entire business process work smoothly. Stocks that are more than optimal will undermine economic performance as a result of asset leasing and storage costs, while smaller stocks will undermine competitive position and endanger security of supply [2].

Accuracy and speed of inventory flow is the main measure of supply chain performance. Stocks in the channel can be replaced by information. If they do not allow each member to distribute their stock information in one channel, at the same time there may be a significant surplus for each member, and there may be a significant deficit in other members. Over the past decades, several changes have taken place in the economy, which has appreciated the role of co-operation.

The competitiveness of the supply chain depends on how the chain members can resolve their conflict of interest. If partners can successfully cooperate in the long term, the chain members can achieve significant economic benefits, which may result in faster delivery levels, shorter lead times, lower purchase costs, and customer satisfaction.

The propagation of the JIT production principles developed in the Japanese automotive industry led to strategic alliances between suppliers and buyers. During the implementation, the first-level suppliers supply complete automation to the car factory (Figure 1). Partners on the second level produce parts for these first-line suppliers. The company that manufactures the assembled finished product has significant rights to the entire chain.

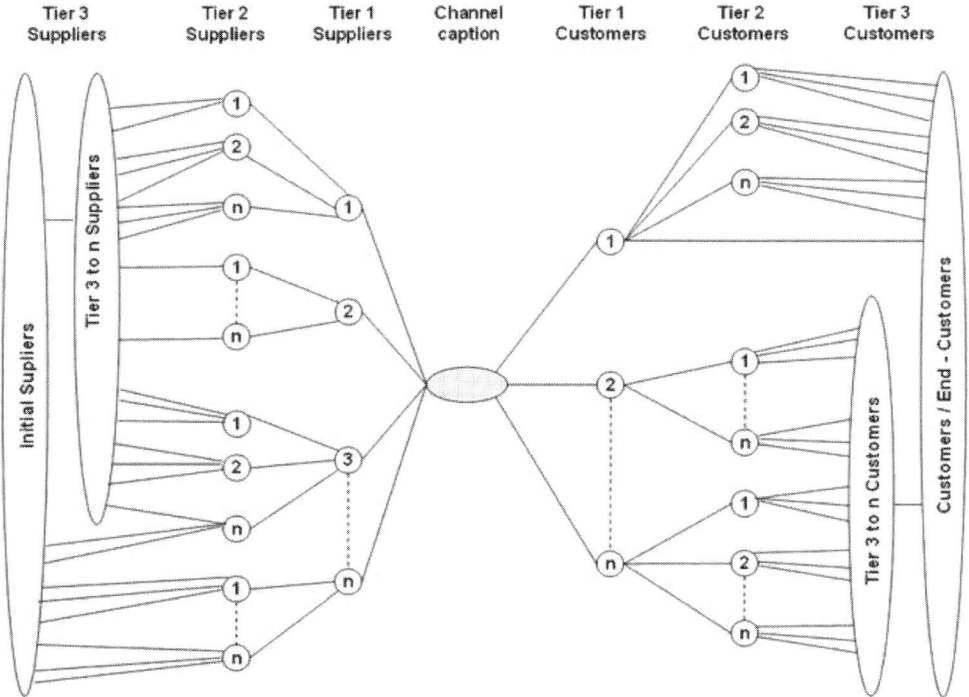

Figure 1: Supply chain network structure [3]

During JIT's production co-operation, the supplier adapts to the customer's processes, and this relationship is characterized by a unilateral warehousing risk of the supplier [4].

Efficient supply chains can only be imagined in a strategic alliance. The strategic alliance between buyers and suppliers is an increasingly noticeable trend. The advantage of the supplier-customer partnership is the sharing of information, which helps the supplier in efficient planning. Creating a strategic alliance also has common benefits as unnecessary order items can be avoided, manual tasks can be automated, and unnecessary control steps can be eliminated from the process.

3   Increase the reliability of the supply chain with horizontal expansion

Reliability of supply chains is of utmost importance for the buyer's competition. In addition to optimizing the supply chain processes, the supply chain structure [5] is becoming more and more important, especially in cooperative forms [6]. While in the case of vertical co-operation, the actors in a series of relationships determine the reliability of the entire supply chain, in the case of horizontal co-operation, parallel partners in the supply chain with each other, even competing with the market, but still cooperating partners. Therefore, during the design and operation of the supply chain, it is necessary to be aware of the development of the reliability attributes that emerge in the organization of the supply chain actors [7].

The probability of a supplier's error-free operating time can be described by exponential distribution based on practical experience:

$$p_o = 1 - e^{-\lambda t} \tag{1}$$

For each of the elements of the supply chain, a so-called κ curve factor can be defined. For this, it is necessary to define the mean time of the average operation time for defective unloading and the average time required for eliminating problems caused by faulty supply. The confusion factor can be described in the following simple context:

$$\kappa = \lambda / \alpha \tag{2}$$

This dependency factor can be used to determine the likelihood of reliability:

$$p_o = 1 / (1 + \kappa) \tag{3}$$

For horizontal co-operation, reliability can be determined by multiplying the reliability of the supply chain operators and the likelihood of operation can be calculated by the following relationship:

$$p_o^{hor} = p_1 p_2 p_3 \dots p_m \tag{4}$$

For vertical co-operation, reliability can be calculated with the following relationship:

$$p_o^{ver} = 1 - (1 - p_1)(1 - p_2)(1 - p_3)(1 - p_m) \tag{5}$$

Business players - particularly in the automotive industry – realises the increasing importance of the vertical supply chain implement horizontal cooperation and thus develop a mixed model of each level. Horizontal co-operation, although at a certain level of the supply chain, results in additional costs, but it increases the reliability of the entire supply chain, thus bringing the supply chain to competitive advantage [8].

Provided an operating cost for a supply chain as an arbitrary or mixed model:

$$C_o^1 = \sum_{i=1}^{m} \sum_{j=1}^{n} c_{i,j} \tag{6}$$

where $c_{i,j}$ is the cost of the j-th cooperating partner at the i-th level of the supply chain.

The resulting reliability of this supply chain should be

$$p_o^1 = p_o^1(\Theta) \tag{7}$$

where Θ is the reliability of the players in the supply chain. If we want to increase the reliability of this supply chain, we either increase the reliability of existing partners or increase the reliability of a given level by involving new partners in horizontal co-operation.

The goal is to involve partners at the given levels for which the increase in operating costs resulting from the excess partners is less than the long-term profit growth resulting from the reliability.

You can reach a vertical level of vertical supply chain by increasing your reliability by horizontal co-operation if the following inequality is met:

$$C_p^2 - C_p^1 > C_o^2 - C_o^1 \tag{8}$$

that is, the gain from the increase in reliability $(C_p^2 - C_p^1)$ is greater than the operating cost of the horizontal expansion.

It is advisable to study three cost components for a simple analysis:

- the cost of horizontal expansion: the higher the credibility element for horizontal co-operation at a level of the vertical supply chain, the greater the maintenance cost of the entire system,
- loss of revenue due to lack of availability of the system: the greater the reliability of the system, the less the loss of production due to the unreliability of the given horizontal level, hence the production company loses less profit,
- revenue, which can be determined as a sum of the cost of horizontal expansion and the loss of revenue loss.

It can be stated that it is not worth to increase the reliability of the system indefinitely, since beyond a certain limit the reliability increase can be solved with costs that can not be offset by a smaller increase in revenue. It can be stated that the increase in the specific maintenance cost of an object resulting in horizontal expansion in a vertical level of a vertical supply chain results in the incorporation of objects with a lower reliability level.

4 Increasing the efficiency of the supply chain through Industry 4.0

The forthcoming Fourth Industrial Revolution or otherwise Industry 4.0 means the accelerating growth of the efficiency of production systems. This includes internet cyber-physical systems, the Internet of Things (IoT) and cloud-based solutions. Based on the cyber-physical systems, integrating the previously unrealistic real and virtual reality, it realizes the new level of organization and regulation of the entire value chain throughout the product life cycle.

The increasingly individualized customer needs are followed by the conceptual design of the cycle and the product, the order, the product development, the production to the delivery, and then the recycling process stations, which also include services related to the product.

The fourth industrial revolution creates what is called a 'smart factory' (Figure 2). In smart factories, cyber-physical systems control the physical processes that create an apparent copy of the physical world and make decentralized decisions. Objects, cyber-physical systems communicate over the Internet and interact with each other and with people in real time.

The fourth industrial revolution can be achieved because today technological innovations and methods have become available, enabling more complex systems to be developed where the entire supply chain can be operated automated. This is based on cyber-physical systems that connect real objects with the processing of information.
Prerequisites for solutions are the creation of smart factories that have independent, shared intelligence and are in close contact with each other [9]. By using detailed and interconnected models of elements, the complex system of these clever components can be optimized for comprehensive goals.

To make this compilation work requires not only the "smart" elements to be networked, but also necessitates their coordination in a "cybernetic space". This is feasible with model-based optimization and development with the use of simulation tools.

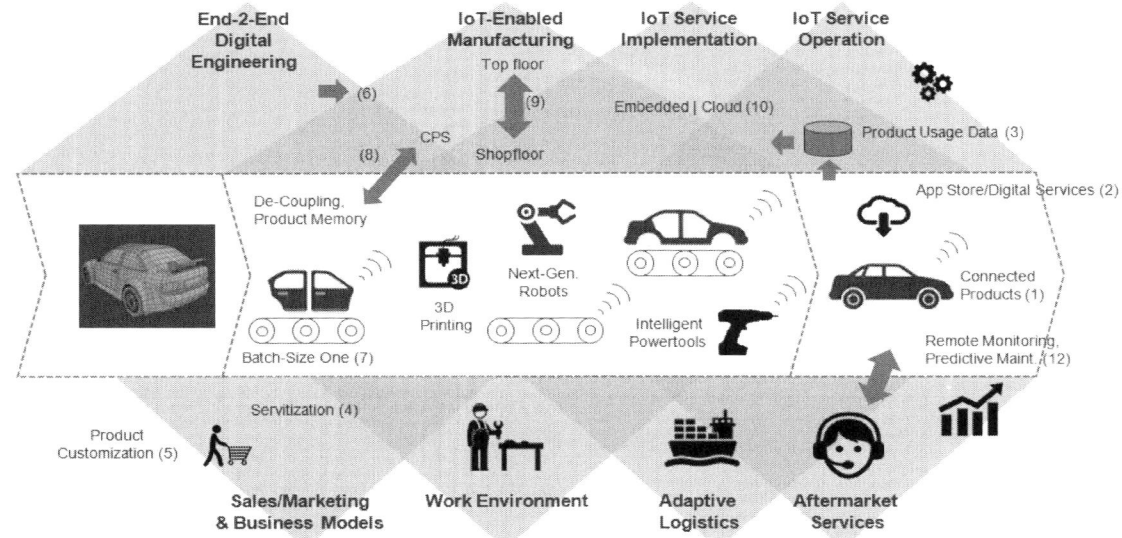

Figure 2: Factory of the Future, Industry 4.0, and the IoT (http://enterprise-iot.org/book/enterprise-iot/part-i/manufacturing/)

Cyber-Physical-Productive Systems (CPPS) do not only incorporate production tools into intelligent networks, but integrate the entire supply chain.

Clever production tools share information about their utilization and status (for example, maintenance) and decide on their own issues independently. Shared information is achieved by using coordination simulation and optimization tools, and optimally also in a completely autonomous way. In order to increase efficiency, utilize capacities, reduce resources, improve quality and reduce lead times, the goal is to coordinate processes.

Smart products have information on their own manufacturing processes, and collect and transmit data from the production and use phases of their life cycle. This makes it possible to create a service offering based on product data and digital modeling of a smart factory.

The development of smart logistics systems, besides sharing information supply chains, is the first to create independent and flexible external and internal logistic solutions. This includes fully automated loading, warehousing, and inter-warehouse (rail, road) handling systems.

However, within Industry 4.0, we have to face many challenges. There is a much greater security risk than cloud computing, cloud computing and frequent sharing of data and information. In particular, the localization and development of non-widespread methods typical of the financial sector, which are less common in industrial production, require the management of product, development, production and market data and the sharing of relevant rights.

5   Industry's 4.0 impact on logistics

Cyber-physical production systems should focus on the entire supply chain management, including maintenance. This requires the capture, archiving of the generated data and the availability of all the elements of the complex network during its operation. This data must cover all aspects of the manufacturing process: production, products, operators, quality change, downtime and management.

Industry 4.0 wants new IT solutions for vertical networking. To create this new solution, there are many things to be available to track and optimize the entire life cycle. For example, supplier data, sensor systems, business applications, and customer data that reflect the business process.

People, machines, and other resources are shaping in a smart factory in a digital model and communicating with cyber-physical systems (CPS).

It assumes and results in high levels of flexibility and transparency in real time, that is, real-time, optimized, new value creation networks. It adapts to the new circumstances of the smart factory environment at all times and optimizes production processes.

The sharing of information on the production systems is assumed to involve the idea, which requires the development of a new business model of cooperation between organizations. This is achieved through integration with suppliers and customers in the value chain.

It should focus primarily on developing new collaboration models with business partners and focusing on customer specific expectations for new business models, which will require new forms of information sharing with customers and suppliers. It is important to mention that continuous communication is required throughout the product's lifecycle, i.e. the management of customer requirements, research and development, procurement, production, and market data.

In the future, supply continues to be the primary task of purchasing and storing, but significant changes are expected in this area:

- Better cooperation with suppliers.
- Localization for faster, environmentally friendly lead times.
- The goods become even more diversified and one-piece orders have to be handled.
- Acquire and store more and more information.
- The demand for traceability covers the entire supply chain.

As a result of localization, they can not only create jobs with greater value, but also reduce complexity, thereby reducing risks in supply chains, lead times and reducing environmental burden by shortening delivery routes.

6   Summary

Automotive companies can maintain their market position by supplying their supply chains effectively, which can be achieved by unlocking the conflict of interests of chain members. The trouble-free operation of the supply chain can be achieved if the chain members create a strategic alliance with the advantage of sharing information. Creating a strategic alliance also has common benefits as unnecessary order items can be avoided, manual tasks can be automated, and unnecessary control steps can be eliminated from the process.

The efficient operation of supply chains can be exploited by exploiting the opportunities offered by the fourth industrial revolution, enabling the entire supply chain to be operated in an automated way. In order to increase efficiency, utilize capacities, reduce resources, improve quality and reduce lead times, the goal is to coordinate processes that can be achieved through Industry 4.0.

7   Acknowledgements

This project has received funding from the EFOP-3.6.1-16-00011 "Younger and Renewing University – Innovative Knowledge City – institutional development of the University of Miskolc aiming at intelligent specialization" project implemented in the framework of the Szechenyi 2020 program and the European Union's Horizon 2020 research and innovation programme under grant agreement No 691942. This research was partially carried out in the framework of the Center of Excellence of Mechatronics and Logistics at the University of Miskolc.

8   References

[1] Cselényi J.; Illés B. (2005): Logistic systems (in Hungarian). Miskolci Egyetemi Kiadó, Miskolc

[2] Nagy, G.; Bányainé Tóth, Á.; Illés, B. (2017): Supply Chain Optimization for Networking Production Companies. In: 10th International Doctoral Student Workshop on Logistics, Magdeburg, pp. 119-124.

[3] Lambert, D. M.; Cooper, M. C.; Pagh, J. D. (1998): Supply chain management: Implementation issues and research opportunities. The International Journal of Logistics Management, Vol. 9 Issue: 2, pp. 1-20.

[4] Szegedi Z. (2012): Supply chain management (in Hungarian). Kossuth Kiadó, Budapest

[5] Te ek, P.; Bányai, T. (2013): Complex design of integrated material flow systems. Advanced Logistic Systems: Theory and Practice 7(1) pp. 105-110.

[6] Bányai, T. (2011): Optimisation of multi-level supply chain of automatised production systems with harmony search algorithm. In: Maria Nowicka-Skowron (szerk.) Proceeding of the II Central European Conference on Logistics 2011. Czestochowa: Politechniki Czestochowskiej, pp. 65-71.

[7] Cselényi J.; Illés B. (2005): Planning and controlling of material flow system (in Hungarian). Miskolci Egyetemi Kiadó, Miskolc

[8] Nagy, G.; Bányainé Tóth, Á.; Illés, B. (2016): The improvement of the efficiency of purchasing by networking operation. Advanced Logistic Systems: Theory and Practice 10(2) pp. 79-90.

[9] Deloitte: Werkpatz 4.0, Basel, 2015.

LOGICAL CHOICE METHOD AMONG POSITIONS IN DOUBLE OPERATION CYCLE OF THE STORAGE VEHICLE

M. Sc. Thanh Dung Cao
Institute of Logistics and Material Handling Systems
Otto von Guericke University Magdeburg, Germany

Univ.-Prof. Dr.-Ing. Hartmut Zadek
Institute of Logistics and Material Handling System
Otto von Guericke University Magdeburg, Germany

1 Abstract

Logical choice among the storage and retrieval positions in the double operation cycle (DOC) is aimed at achieving the best efficiencies of the energy need and the throughput of an automated storage and retrieval vehicle (SRV) in automatic small parts warehouse (ASPW). The main characteristic of the throughput is the vehicle's moving time.

The method of choosing the pairs of DOC is derived from simple choices with small numbers of the storage and retrieval compartments. After that, it is extended to the entire system to achieve the best efficiencies of the energy need and the moving time.

A simulation model of DOC is accordingly developed from the simulation model of the single operation cycle (SOC) [1] to simulate the kinematic parameters, the power, the energy need and the moving time of the driving unit and the lifting unit. The energy recovery is included in this simulation to describe the same as the experimental model.

The efficiencies of the logical choice method among the storage and retrieval positions to reduce the energy need and the moving time in DOC of the system are shown out by the results of the simulation model and the specific analyses.

2 Introduction

Energy efficiency is at the heart of the EU's Europe 2020 Strategy for smart, sustainable and inclusive growth and of the transition to a resource efficient economy. Energy efficiency is one of the most cost effective ways to enhance security of energy supply, and to reduce emissions of greenhouse gases and other pollutants [2]. Besides, the national sustainability strategies of the German Federal Government have been issued since 2002 [3] and one of the areas of the special interest is the goal of reducing the primary energy use in subsequent years about 20% by 2020 and about 50% by 2050 compared to 2008. On the other hand, greenhouse gas emissions are required to reduce about 40% by 2020 and from 80% to 95% by 2050 compared to 1990 [4,5]. Nowadays, Logistics plays an important role in terms of the competitiveness and the economy for manufacturing companies as well as commercial enterprises [6]. In Logistics, Intralogistics is calculated for about 24% of energy consumption [7]. It contains potentials with great impact to the economy and the environment so it is a priority field that should be considered.

ASPW is increasingly used in Intralogistics to guarantee the high profitability due to its suitable storage characteristics and optimal existing storage space. ASPW is designed for lightweight and small units up to 50 kg that are stored in totes, cardboard boxes or on trays depending on the goods type, the weight, the requirements of the throughput and the application field. The height of ASPW is usually below 18 m [8].

SRV is one of the main components of ASPW. It is a handling and lifting machine limited on the rails and it travels within and out of the aisles for the storage and retrieval of unit loads and/or for order picking or similar duties. This vehicle shall either include lifting means and/or lateral handling facilities. Control of vehicles may range from manual to fully automatic [9].

Saving the energy need and increasing the throughput of SRV during the storage and retrieval processes are major factors affecting the productivity and cost of goods. As a result, it is more competitive for enterprises when the transportation cost in the storage and retrieval processes is reduced and the productivity is increased.

In the past, the method to optimize the energy need and the throughput in SOC was shown out [1]. Besides, the average values of the throughput and the energy need of the different storage strategies in DOC were determined when the kinematic parameters of SRV changed [10], etc. However, the logical choice method among the storage and retrieval positions of DOC in the system has not been pointed out to reduce the energy need and increase the throughput of SRV. This method is shown in detail as below.

3 Simulation model of DOC

To clarify the effectiveness of the logical choice method, the dimensions of the considering system is extended from the experimental model at the Institute of Logistics and Material Handling Systems (ILM) at the Otto-von-Guericke University Magdeburg [1]. It must ensure that the experimental coefficients are not affected by the simulation model to fit the reality. These coefficients depend on the structure of SRV. It means that the storage and retrieval system cannot change vertically when these coefficients are applied to the simulation model. The reason is when the height of the system changes, the structure of SRV has to change to suit the system and the coefficients are incorrect in this case and then the experimental coefficients must be redefined. The system can be changed horizontally, then the structure of SRV is constant and the simulation results are similar to the reality. The operation behavior of a storage and retrieval system with length of 22.5 m and height of 7.5 m including the engine foundation and the top of SRV is numerically simulated in this study. There are two identical racks on the sides, of which SRV moves in the middle. Each rack has 45 compartments horizontally and 21 compartments vertically. The total of two racks is 1890 compartments. Each compartment has a width of 0.5 m and a height of 0.31 m. The input/output point (I/O) is located in front of the rack at the height of the second row. The energy recovery is also included in this system. The compartments denoted from a1 to a4 are illustrated in Fig. 1.

In DOC, SRV moves from input point (I) to the storage compartment, then to the retrieval compartment and ends at output point (O). The computational formulas of the kinematic parameters, the power, the energy need, the working cases and the experimental coefficients of SRV in every distance of DOC are taken as in the simulation model of SOC [1].

The input parameters of the simulation model include the size and the positions of the storage and retrieval compartments; the requirement of the speed, the acceleration and the jolt of the driving unit and the lifting unit; the mass of the unit load; the parameters of SRV consist of the mass of SRV and the lifting unit, the diameter of the wheel, the diameter of the wheel gudgeon; the experimental coefficients of the driving unit and the lifting unit.

The results of the simulation model consist of simulating the moving and lifting distance, the speed, the acceleration and the consumption power by the driving and lifting time from the input point to the storage compartment and then to the retrieval compartment and to the output point; simulating the recovery power of the driving unit when SRV is deceleration or braking; simulating the recovery power of the lifting unit when it lowers the unit load; determining the total energy need and the total moving time when SRV moves from input point to output point with every input value of the speed and the acceleration.

4 Theoretical basis for determining logical choice method among positions of DOC

To determine the logical choice method among positions, the energy need and the moving time of SRV in DOC are compared one by one in some simple cases. After that, the rule for the logical choice is defined and then extended to the entire system to achieve the best efficiencies of the energy need and the moving time based on the inductive approach.

When a storage position and a retrieval position are chosen in a pair of DOC: the farther from the I/O point the storage position and the retrieval position are and the nearer to each other by horizontal direction they are, the higher the saving efficiencies of the energy need and the moving time in DOC are.

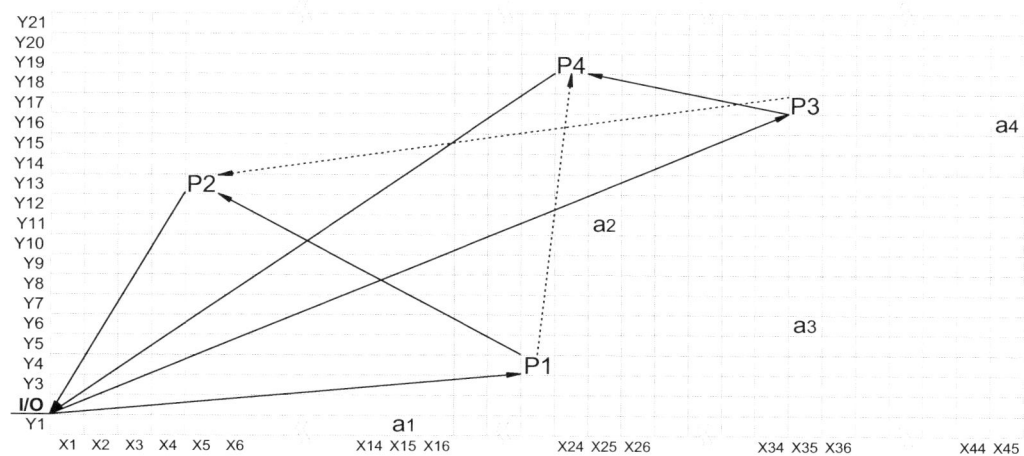

Figure 1: The specific positions of the storage and retrieval system.

In the next step, it is assumed that there are four positions selected for any two pairs of DOC. P1 and P3 are two storage positions, P2 and P4 are two retrieval positions accordingly (Fig. 1).
When these positions are changed, the best way to choose the suitable pairs of DOC can be found by analyzing some cases.
The energy need and the working time of the load handling device in every DOC are the same and then in order to simplify, they are not included in the next analysis.
Four positions can be selected in two ways as below:
The first way is two pairs P1P2 and P3P4 and the second way is two pairs P1P4 and P3P2.
The total moving time of the driving unit and the lifting unit in the first way t_{TDC1} is calculated by Equation (1) and in the second way t_{TDC2} by Equation (2); the difference moving time between two options is shown by Equation (3). The energy need of the driving unit and the lifting unit in the first way E_{TDC1} is presented by Equation (4) and in the second way E_{TDC2} by Equation (5); the difference energy need between two options is shown by Equation (6).

t_{dri} : The driving time

t_{lift} : The lifting time

E_{dri} : The energy need of the driving unit

$E_{dri\,recu}$: The energy recovery of the driving unit

E_{lift} : The energy need of the lifting unit

$E_{lift\,recu}$: The energy recovery of the lifting unit

$$t_{Ii} = Max\left\{t_{dri\,IPi}; t_{lift\,IPi}\right\}$$

$$t_{ij} = Max\left\{t_{dri\,PiPj}; t_{lift\,PiPj}\right\}$$

$$t_{iO} = Max\left\{t_{dri\,PiO}; t_{lift\,PiO}\right\}$$

$$E_{DIi} = E_{dri\,IPi} - \left|E_{dri\,reco\,IPi}\right|$$

$$E_{Dij} = E_{dri\,PiPj} - \left|E_{dri\,reco\,PiPj}\right|$$

$$E_{DiO} = E_{dr.\,PiO} - \left|E_{dri\,reco\,PiO}\right|$$

$$E_{LIi} = E_{lift\,IPi} - \left|E_{lift\,reco\,IPi}\right|$$

$$E_{Lij} = E_{lift\,PiPj} - \left|E_{lift\,reco\,PiPj}\right|$$

$$E_{LiO} = E_{lift\,PiO} - \left|E_{lift\,reco\,PiO}\right|$$

$$t_{TDC1} = t_{I1} + t_{I3} + t_{2O} + t_{4O} + t_{12} + t_{34} \tag{1}$$

$$t_{TDC2} = t_{I1} + t_{I3} + t_{2O} + t_{4O} + t_{14} + t_{32} \tag{2}$$

$$t_{TDC2} - t_{TDC1} = t_{14} + t_{32} - (t_{12} + t_{34}) \tag{3}$$

$$E_{TDC1} = E_{DI1} + E_{DI3} + E_{D2O} + E_{D4O} + E_{LI1} + E_{LI3} \tag{4}$$
$$+ E_{L2O} + E_{L4O} + E_{D12} + E_{D34} + E_{L12} + E_{L34}$$

$$E_{TDC2} = E_{DI1} + E_{DI3} + E_{D2O} + E_{D4O} + E_{LI1} + E_{LI3} \tag{5}$$
$$+ E_{L2O} + E_{L4O} + E_{D14} + E_{D32} + E_{L14} + E_{L32}$$

$$E_{TDC2} - E_{TDC1} = E_{D14} + E_{D32} - E_{D12} - E_{D34} \tag{6}$$
$$+ E_{L14} + E_{L32} - E_{L12} - E_{L34}$$

The logical choice method for two pairs of DOC by selecting any two storage positions and two retrieval positions is an important step in finding the suitable rule for the entire system. Therefore, it is considered in detail.

5 Efficiency of logical choice for two pairs of DOC in the vertical

The vertical is considered in priority to find some remarks for the logical choice method between two pairs of DOC.
Based on Equation (6), the different energy need of the lifting unit E_{Ldif} is considered first by Equation (7). In this case, only the height of the positions is considered. Due to P1 and P3 are the storage positions, P2 and P4 are the retrieval positions, it is assumed that the storage position P1 is lower than P3 and the retrieval position P2 is lower than P4.

$$E_{Ldif} = \bar{E}_{L14} + E_{L32} - E_{L12} - E_{L34} \tag{7}$$

To determine the different energy need of the lifting unit E_{Ldif} in any way, some cases are compared when the heights of the storage and retrieval positions are changed (Fig. 2).

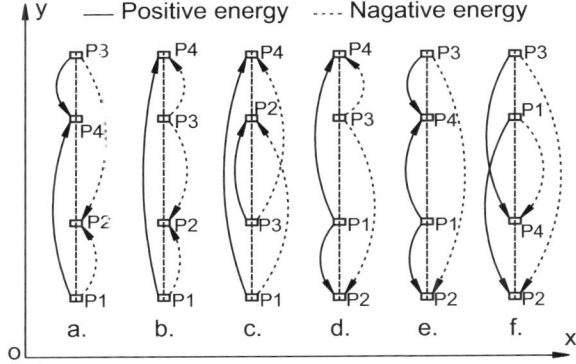

Figure 2: Cases of storage and retrieval positions vertically

The different energy need of the lifting unit is determined by Fig. 2a and Eq. (8) when P2 and P4 are between P1 and P3; by Fig. 2b and Eq. (9) when P4 is at the top and P2 is between P1 and P3; by Fig. 2c and Eq. (10) when P1 and P3 are under P2 and P4; by Fig. 2d and Eq. (11) when P1 and P3 are between P2 and P4; by Fig. 2e and Eq. (12) when P3 is at the top and P1 is between P2 and P4; by Fig. 2f and Eq. (13) when P1 and P3 are above P2 and P4.

$$E_{Ldif\,a} = \left|E_{L14}\right| - \left|E_{L32}\right| - \left|E_{L12}\right| + \left|E_{L34}\right| \tag{8}$$

$$E_{Ldif\,b} = \left|E_{L14}\right| - \left|E_{L32}\right| - \left|E_{L12}\right| - \left|E_{L34}\right| \tag{9}$$

$$E_{Ldif\,c} = \left|E_{L14}\right| + \left|E_{L32}\right| - \left|E_{L12}\right| - \left|E_{L34}\right| \tag{10}$$

$$E_{Ldif\,d} = \left|E_{L14}\right| - \left|E_{L32}\right| + \left|E_{L12}\right| - \left|E_{L34}\right| \tag{11}$$

$$E_{Ldif\,e} = \left|E_{L14}\right| - \left|E_{L32}\right| + \left|E_{L12}\right| + \left|E_{L34}\right| \tag{12}$$

$$E_{Ldif\,e} = -\left|E_{L14}\right| - \left|E_{L32}\right| + \left|E_{L12}\right| + \left|E_{L34}\right| \tag{13}$$

Positive energy and negative energy in all cases from Equation (8) to Equation (13) are always performed on total equal altitudes, (see Fig. 2). Besides, the height of the system is average and the energy recovery rate of the lifting unit based on the experimental results is about 70% its energy consumption on every height and every kinematic parameter and then E_{Ldif} in every case

is small and affects a little bit to total energy need in Equation (6). Therefore, the next analyses in Equation (6) only focus on the energy need of the driving unit when the selected ways are compared.

The speed and the acceleration of the lifting unit are quite big ($v_{linp} = v_{lmax} = 4\,\text{m/s}$ and $a_{linp} = a_{lmax} = 4\,\text{m/s}^2$) and the maximum lifting height is average. As a result, in most cases, the height difference among the positions is not much and when SRV moves from the storage position to the retrieval position, the lifting time is less than the driving time. Only a few cases, the difference in height among the positions is quite big compared to the difference in the driving distance among positions, then the lifting time is a little bit bigger than the driving time and it does not significantly affect to the total working time of SRV. The next analyses of Equation (3) only consequently focus on the driving time when the selected ways are compared to each other.

Therefore, the height of the positions can be ignored when the energy need and the moving time one by one in the cases of DOC are compared to each other.

6 Efficiency of logical choice for two pairs of DOC in the horizontal

The height of the positions is ignored when the positions are chosen to achieve the high efficiencies of the energy need and the moving time. Therefore, only the horizontal of the positions is considered.

It is assumed that the storage position P1 is always nearer than P3 and the retrieval position P2 is always nearer than P4 horizontally. Some cases are indicated to determine the logical choice method among the positions when four positions are changed, see Fig 3 and some examples on Table 1.

The cases are considered as below to determine logical choice method among the positions:

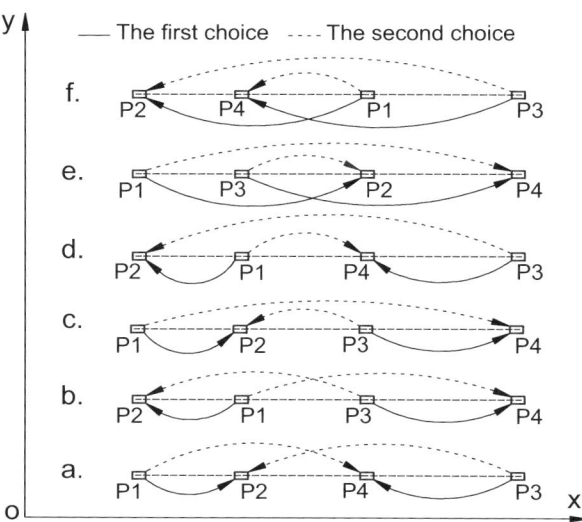

Figure 3: Cases of storage and retrieval positions horizontally

Case a: when P2 and P4 are in between P1 and P3 horizontally (Fig. 3a)
The farthest storage position P3 is chosen in a pair with the farthest retrieval position P4 and P1 is chosen in a pair with P2. After that, the saving efficiencies of the energy need and the moving time in the first way (P1P2 & P3P4) are better than in the second way (P1P4 & P3P2). The reason is that from Fig. 3a, the driving distance from P1 to P4 (s_{d14}) is bigger than the one from P1 to P2 (s_{d12}) and the driving distance from P3 to P2 (s_{d32}) is bigger than the one from P3 to P4 (s_{d34}) $(s_{d14} > s_{d12}$ and $s_{d32} > s_{d34})$. Therefore, from Equation (3) and (6), it is easy to know $t_{TDC2} - t_{TDC1} > 0$ and $E_{TDC2} - E_{TDC1} > 0$. For example: Table 1 Case a, the moving time by the second choice (a1a3 & a4a2) is higher 8.39% than the one by the first choice (a1a2 & a4a3) $(PMT(a) = 8.39\%)$ and the energy need by the second choice is higher 13.59% than the one of the first choice $(PEN(a) = 13.59\%)$.

Case b: when P1 and P3 are in between P2 and P4 horizontally (Fig. 3b)
The farthest storage position P3 is chosen in a pair with the farthest retrieval position P4 and P1 is chosen in a pair with P2. After that, the efficiencies of the energy need and the moving time in the first way (P1P2 & P3P4) are better than in the second way (P1P4 & P3P2). The reason is that from Fig. 3b, $s_{d14} > s_{d34}$ and $s_{d32} > s_{d12}$. Therefore, from Equation (3) and (6), it is easy to know $t_{TDC2} - t_{TDC1} > 0$ and $E_{TDC2} - E_{TDC1} > 0$. For example: Table 1 Case b, $PMT(b) = 8.39\%$ and $PEN(b) = 13.6\%$.

Case	Storage positions	Retrieval positions	Combinations of two pairs	Total driving and lifting energy need (kWs)	PEN (%)	Total driving and lifting time (s)	PMT (%)
a	a1 & a4	a2 & a3	a1a2 & a4a3	127.65	13.59	28.6	8.39
			a1a3 & a4a2	145		31	
b	a2 & a3	a1 & a4	a2a1 & a3a4	127.51	13.60	28.6	8.39
			a2a4 & a3a1	144.85		31	
c	a1 & a3	a2 & a4	a1a2 & a3a4	126.36	14.93	28.6	8.57
			a1a4 & a3a2	145.22		31.05	
d	a2 & a4	a1 & a3	a2a1 & a4a3	128.8	15.05	28.6	8.57
			a2a3 & a4a1	148.19		31.05	
e	a1 & a2	a3 & a4	a1a3 & a2a4	144.28	1.35	31	0.16
			a1a4 & a2a3	146.23		31.05	
f	a3 & a4	a1 & a2	a3a1 & a4a2	145.57	1.11	31	0.16
			a3a2 & a4a1	147.18		31.05	

Table 1: Some examples to determine logical choice method among four positions, when the driving unit speed is 4 m/s and the driving unit acceleration is 3 m/s².

Case c: when P2 is in between P1 and P3 horizontally, P4 is the farthest (Fig. 3c)
The farthest storage position P3 is chosen in a pair with the farthest retrieval position P4 and P1 is chosen in a pair with P2. After that, the efficiencies of the energy need and the moving time in the first way (P1P2 & P3P4) are better than in the second way (P1P4 & P3P2).
The reason is that from Fig. 3c, $s_{d14} > s_{d12} + s_{d34}$.
Therefore, from Equation (3) and (6), $t_{TDC2} - t_{TDC1} > 0$ and $E_{TDC2} - E_{TDC1} > 0$. For example: Table 1 Case c, $PMT(c) = 8.57\%$ and $PEN(c) = 14.93\%$.
It means that, the efficiencies of the energy need and the moving time of the first choice (P1P2 & P3P4) in this case are higher than the ones of the second choice.

Case d: when P1 is in between P2 and P4 horizontally, P3 is the farthest (Fig. 3d)
The farthest storage position P3 is chosen in a pair with the farthest retrieval position P4 and P1 is chosen in a pair with P2. After that, the efficiencies of the energy need and the moving time in the first way (P1P2 & P3P4) are better than in the second way (P1P4 & P3P2). The reason is that from Fig. 3d, $s_{d32} > s_{d12} + s_{d34}$.
Therefore, from Equation (3) and (6), it is easy to know $t_{TDC2} - t_{TDC1} > 0$ and $E_{TDC2} - E_{TDC1} > 0$. For example: Table 1 Case d, $PMT(d) = 8.57\%$ and $PEN(d) = 15.05\%$.

Case e,f: when P1 and P3 are on the same side, P2 and P4 are on the other side (Fig. 3e,f)
The results of the energy need and the moving time of the two options are nearly equal. The reason is that from Fig. 3e,f, $s_{d14} + s_{d32} = s_{d12} + s_{d34}$.
Therefore, from Equation (3) and (6), $t_{TDC2} \approx t_{TDC1}$ and $E_{TDC2} \approx E_{TDC1}$. It is only slightly different due to the different starting and braking distances or due to the influence of the energy need and the lifting time of the lifting unit by the height of the different compartments. For example: Table 1 Case e, $PMT(e) = 0.16\%$ and $PEN(e) = 1.35\%$; Table 1 Case f, $PMT(f) = 0.16\%$ and $PEN(f) = 1.11\%$.
From 6 above cases, the general method to choose two pairs of DOC from any four positions to achieve the best efficiencies of the energy need and the moving time is shown that the farthest storage position is chosen in a pair with the farthest retrieval position and the remaining storage position is chosen in a pair with the remaining retrieval position.

7 Summary

When a storage position and a retrieval position are chosen in a pair of DOC: the farther from the I/O point the storage position and the retrieval position are and the nearer to each other by horizontal direction they are, the higher the saving efficiencies of the energy need and the moving time in DOC are.

When two storage positions and two retrieval positions are chosen in any two pairs of DOC, the general method to choose two pairs of DOC to achieve the best efficiencies of the energy need and the moving time is shown: the farthest storage position is chosen in a pair with the farthest retrieval position and the remaining storage position is chosen in a pair with the remaining retrieval position.

From two above results, the method to choose the pairs of DOC can be extended to the entire system based on the inductive approach to achieve the best efficiencies of the energy need and the moving time. This method is shown that: *the farthest storage compartment and the farthest retrieval compartment by horizontal direction are always selected as the first pair and then the second pair is selected at the second farthest storage and retrieval compartments and so on. Finally, the storage and retrieval compartments at the nearest positions to I/O point by horizontal direction are chosen*. The system achieves the best efficiencies of the energy need and the moving time in DOC respectively.

This method is only applied when SRV has an energy recovery system and the system's height is just average. If the system is high and then the energy efficiency of this method is still suitable due to the energy recovery rate of the lifting unit is about 70% the energy consumption of the lifting unit on every height. However, the lifting time is quite big and it directly affects to the total working time of SRV then the saving efficiency of the moving time is not high. In this case, depending on the height of the system, it is possible to divide the system into zones of height and this method is applied in each zone. After that, the saving efficiencies of the energy need and the moving time can are still suitable.

8 References

[1] Cao, T. D.; Zadek, H. (2017): Controlling storage vehicle in distances for only the startup, braking phases and optimizing the energy need. In: 10th International Doctoral Students Workshop on Logistics. Otto-von-Guericke-University Magdeburg, P. 39-44.

[2] European Commission (2011): Communication from the commission to the European parliament, the council, the European economic and social committee and the committee of the regions - Energy Efficiency Plan 2011, P. 2.

[3] Bundesregierung Deutschlands (2002): Perspektiven für Deutschland - Unsere Strategie für eine nachhaltige Entwicklung. Berlin.

[4] Bundesregierung Deutschlands (2016): Deutsche Nachhaltigkeitsstrategie. Berlin, P. 37-39.

[5] Bundesregierung Deutschlands (2012): Nationale Nachhaltigkeitsstrategie - Fortschrittsbericht. Berlin, P. 29.

[6] Fusko, M.; Rakyta, M.Manlig, F. (2017): Reducing of Intralogistics Costs of Spare Parts and Material of Implementation Digitization in Maintenance. Procedia Engineering 192:213-218.

[7] Altintas, O.; Avsar, C.; Klumpp, M. (2010): Change to Green in Intralogistics. In: The 2010 European Simulation and Modelling Conference. Hasselt University, Oostende, P. 373-377.

[8] Günthner, W. A.; Atz, T.; Ulbrich, A. (2011): Forschungsbericht zum Projekt integrierte lagersystemplanung. TU München, Garching bei München, P. 8-10.

[9] European Materials Handling Federation (2016): FEM 9.101, Guideline / Terminology - Storage and Retrieval Machines - Definitions. Frankfurt, P. 6.

[10] Schulz, R. (2014): Untersuchung und Ableitung geeigneter Lagerbetriebsstrategien zur Verringerung des Energiebedarfs von Regalbediengeräten. Otto-von-Guericke-University Magdeburg, P. 139-171.

LOGISTICAL RISKS OF INDUSTRIAL COOPERATION OF THE VIRTUAL ENTERPRISE WITHIN B2B RELATIONS

Anastasiia Bezsmertna
Department of Theoretical Mechanics and Engineering and Robotic Systems/ Aircraft Engines Faculty
National Aerospace University "Kharkiv Aviation Institute"

Nataliya Rudenko
Department of Theoretical Mechanics and Engineering and Robotic Systems/ Aircraft Engines Faculty
National Aerospace University "Kharkiv Aviation Institute"

Olga Nefedkina
Department of Foreign languages/Humanities Faculty
National Aerospace University "Kharkiv Aviation Institute"

1 Formulation of the problem

In the modern "information" economy, one of the main factors of competitive ability is the speed of reaction to operational market changes. The development of information technologies, changes in the competitive situation in the market and the increasingly narrow specialization in the spheres of production and services cause the emergence of new business forms. One such form is virtual enterprises. The virtual enterprise is based on the single organizational formation, technological and information environment through the temporary pooling of various enterprises resources. On the basis of the operational coordination of the resources using, enterprises are able to produce the final product or service quickly and at minimal cost. Industrial cooperation actively develops within B2B relations when the large industrial enterprises enter the close cooperation relations with the small and medium-sized enterprises making for them certain details and accessories. Industrial cooperation in modern conditions is inherently a logistical system. Logistic system in any practical implementation is from the process of moving cargo to the processes of commodity circulation in the market space. It includes a variety of heterogeneous elements, the functioning of which is influenced by various factors and involves a certain risk.

One of the principles of logistics is reliability, and at the micro- and macro-logistical levels. This means that the risk of the functioning of the logistics system should be minimized or even neutralized [1]. Accounting of risk factors has the features at all stages life cycle of an industrial product. With regard to the decision is making phases on cooperation and the choice of partners, organizational risk issues are most relevant. Consideration of operational risks is particular importance at the stages of work planning and implementation. When building a cooperative business strategy, enterprises inevitably face is not only new opportunities, but it is also new potential dangers. A virtual enterprise is located in a zone of high risk, since it is real both for the traditional risks of an industrial enterprise, risks associated with the virtual form of the organization.

2 Analysis of recent research and publications

In the 1980s, the main directions for improving the activities of enterprises were total quality management and the application of minimalist strategies aimed at optimal management of various resources. In the 90s, the main slogan was the principles of business process reengineering, aimed at moving from functional units to business processes consisting of autonomous groups focused on better satisfying the interests of customers. By the end of the 1990s and the beginning of the 21st century, the key topic was the transition to virtual and network principles of enterprise organization [1, 2]. Virtual enterprises are one of the newest organizational forms of management. Their emergence is associated with globalization and the development of modern markets oriented towards sustainable relations with consumers, as well as the growing importance degree application of new information and communication technologies.

In practice, there is a multitude of "virtualized" organizations to one degree or another. It should be noted that quite often under the virtual organizations in the production understanding some form interaction organization of enterprises in logistics. The concept of management supply chain includes the planning and management of all activities related to the selection of suppliers, logistics and processing, as well as all operations for the management of logistics. It is important to note that this concept also includes coordination and cooperation processes among partners in the distribution channel, among which there may be suppliers, resellers, third-party service providers and customers.

In general, virtual organizational forms are dynamic network associations of people, collectives and enterprises [1]. One of the most important advantages of such an organization is a sharp reduction in the amount of seed capital for the establishment of a new business, since most of the necessary resources will be contracted and paid as services are provided. Another advantage is a significant reduction in the time needed to prepare for the next project [3].

3 Purpose of the study

The study is devoted to the analysis and assessment of logistics risks of industrial cooperation of a virtual enterprise within the framework of B2B relationships. In the course of the study, it is planned to compile an exhaustive list of risks for the functioning of the virtual enterprise. In this case, the concept of the product life cycle will be used.

4 Materials and research results

Schematically, the mechanism of the virtual enterprise is shown in Figure 1.

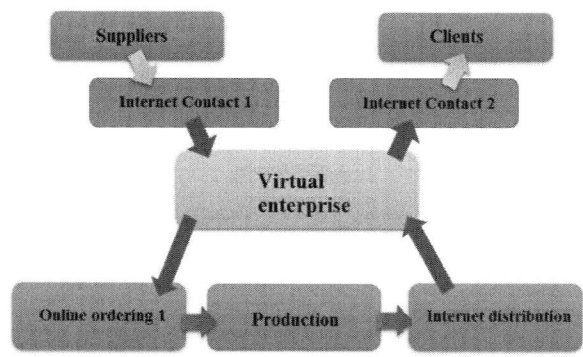

Figure 1: The mechanism of the virtual enterprise

The supplier carries out entrepreneurial activities in accordance with the terms of the concluded supply contract, which is one of the types of the contract of sale. In accordance with the supply agreement, the supplier undertakes to transfer the goods produced or purchased by the buyer within a specified time or period for use in business or for other purposes not related to personal, family, domestic or other similar use.

The customer is a person (physical or legal) who is interested in executing the work, rendering services to them, or purchasing a product from the seller (in a broad sense). Sometimes it is assumed that the order is made, but not necessarily.

Internet contact 1, Internet contact 2, Internet order, Internet distribution - means of communication and interrelations between the elements of the system.

Orders from the client come through the means of communication (telephone, fax, e-mail, etc.) to the marketing department. Then we select the orders for the batch. The chief engineer selects a package of design and technological documentation. Also, his duties include the selection of production agents and the placement of orders for the manufacture of nodes.

The chief engineer makes orders for components for the production of units and final assembly. Components for the production of nodes from suppliers directly go to a manufacturing enterprise and components for assembly - to a virtual enterprise. Delivery of ready assemblies from production agents to assembly is carried out by a logistics company. Further in the technical department of the virtual enterprise, the finished product is assembled. Quality control and pre-sale preparation is carried out by the commercial director.

The list of risks for the functioning of a virtual enterprise will be compiled on the basis concept of the product life cycle. The product life cycle includes a number of steps, from the birth idea of a new product to its disposal at the end of its life [4].

In Ramus Educational environment, we construct a contextual diagram of the product life cycle. At the entrance we receive the order, accordingly on an output we receive performance of the order. Managing impact is the time it takes to produce one unit of output and the planned risks that can occur at different stages of the life cycle. The work execution mechanisms are the chief engineer, commercial director, design office, technology department, transport service and marketing group. Then the context diagram is decomposed into 6 stages. Input and output information, control action and work execution mechanisms remain the same as in the context diagram.

Marketing. The main task of marketing is to understand the needs and requirements of each market and choose those that their company can service better than others. At the output of this stage, we receive a formed order. Managing impact is time. The mechanism of performance of works is a marketing group.

Product design. At the entrance of this stage we receive the formed order, at the output is a set of design and technological documentation. Managing impact is time. The work execution mechanism is the design bureau and chief engineer.

Preparation of production. At the entrance of this stage we get a set of design and technological documentation, at the output is the prepared production. The managing impact is the time and the resulting set of design and technological documentation. The mechanism of work execution is the technological department and the chief engineer.

Production. At the entrance of this stage we get the prepared production, at the output is the finished batch of products. Managing impact is time and planned risks. The mechanism of work execution is the technological department, the transport service and the chief engineer.

Realization. At the entrance of this stage, we get a finished batch of products, at the output is delivery and installation of the product. Managing impact is the planned risks. The mechanism of work execution is the marketing group, the transport service and the chief engineer.

Exploitation. At the entrance of this stage we receive the delivered and installed product, at the output is the execution of the order. Managing impact is the planned risks. The work execution mechanism is the technology department and the commercial director

On the basis of the resulting product decomposition diagram, we compile a list of the risks of the virtual enterprise that arise at each stage of the product life cycle, and determine the reasons for their appearance (Figure 2).

We will assess the logistics risks of industrial cooperation of a virtual enterprise using the number of priority risks. The number of priority risk is the risk assessment method used in the analysis of species and consequences of failures, which is determined by formula

$$NPR = S \cdot O \cdot D \qquad (1)$$

where S – is an indicator of the significance or criticality of failure;
O – is an indicator of the probability or frequency of the cause of failure;
D – is indcator of the probability of finding a defect or error

10	High	Potential failure can lead to a risk of loss of health and life of the worker, customer or supplier.
9		A potential failure affects the safe functioning and / or leads to a violation of the law.
8		Failure leads to loss of functionality of the entire product.
7	Average	Refusal leads to extreme dissatisfaction of the customer / buyer.
6		Failure entails a significant loss of functionality of the entire product.
5		Failure results in a partial loss of functionality.
4		Failure does not affect the functionality of the product, but leads to dissatisfaction of the customer / buyer.
3	Low	Failure affects functionality, but can be seen by an ordinary user.
2		Failure does not result in significant consequences and will not be noticed by an ordinary user.
1		There is no visible defect.

Table 1: Criterion for calculating "S"

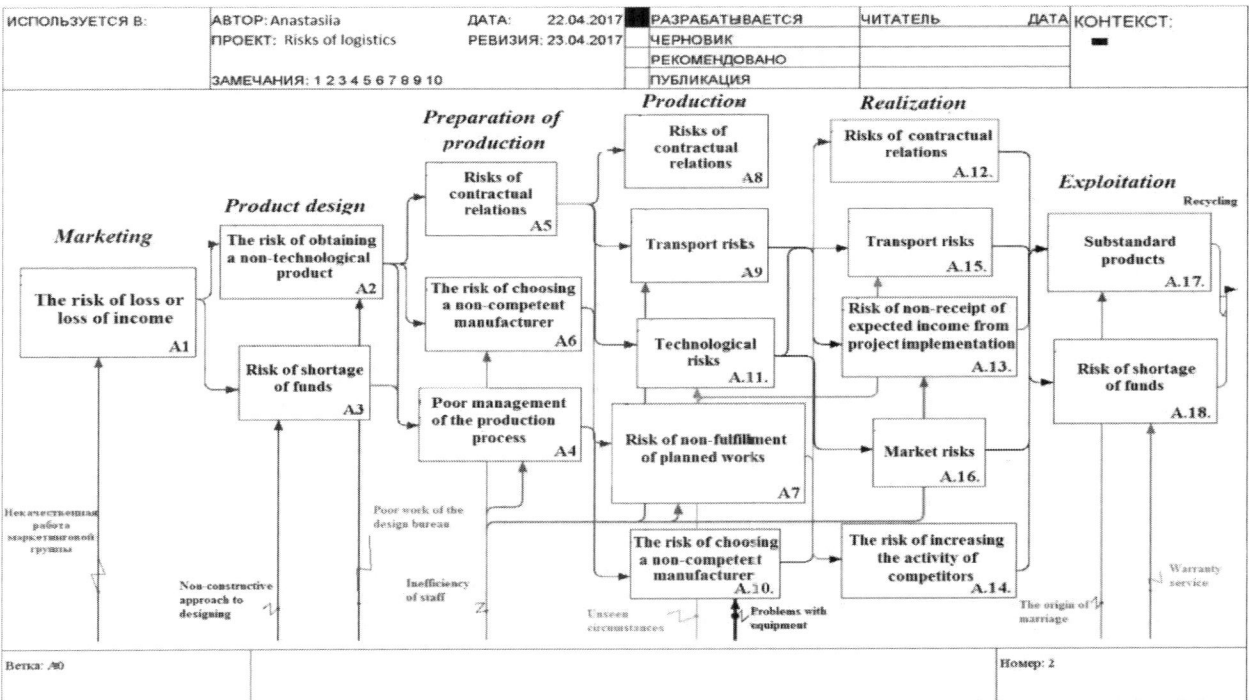

Figure 2: Risks arising at the stages of the product life cycle

Rank	Probability		Possible fractions of failure		C_{PK}	P_{PK}
10	Very high	Failure is almost inevitable	More often 1 time a day	300000 PPM	<0,33	<0,55
9	High	The probability of failure is similar to the probability of their failure Serial failures	1 time in 3-4 days	100000 PPM	≈0,33	≥0,55
8			Once a week	50000 PPM	≈0,67	≥0,78
7			1 time per month	10000 PPM	≈0,83	≥0,86
6	Average	Frequent failures	1 time in 3-4 months	1000 PPM	≈1	≥0,94
5		Random Failures	1 time in half a year	500 PPM	≈1,17	≥1
4		Unstable failures	1 time per year	100 PPM	≈1,33	≥1,1
3	Low	Rare Failures	Once in 2-3 years	50 PPM	≈1,67	≥1,2
2		There may be occasional failures	1 every 3-5 years	10 PPM	≈2	≥1,3
1		Failure is unlikely	Less than 1 time in 5 years	< 2 PPM	>2	≥1,67

Table 2: Criterion for calculating "O"

Rank	Criterion	Definition
10	Impossibility of detection	The presence of a defect is not verified or can not be found
9	The defect most likely will not be detected	The product is selectively checked and evaluated based on the acceptable level of quality / marriage
8	High probability of non-detection	The entire product is checked visually and is evaluated based on the absence of defects
7	There is a probability of detection	The product is visually inspected during the manufacturing process
6	Very low probability of detection	The product is visually examined using a gold sample
5	Low probability of detection	The process is controlled statically, but is evaluated

		outside the line
4	Average probability of detection	The process is statically controlled and evaluated directly during production
3	High probability of detection	The process is statically controlled
2	Almost complete probability of detection	All products are checked automatically
1	The probability of detection is 100%	All products are checked automatically and the defect can not be missed

Table 3: Criterion for calculating "D"

Based on the data on the calculation of all criteria, we perform an assessment of all factors. As a result, we determine the priority number of risks.

Main factors	Composition of factors	Significance S	Probability of occurrence O	Probability of detection D	NPR
Suppliers	Reliability	10	6	7	420
	Fare	8	6	7	336
	Certification	7	4	1	28
	Mode of supply	5	5	2	50
	Price	9	5	1	45
	Terms of payment	8	5	1	40
External environment	Legislation	6	4	3	72
	State of the economy	7	9	7	441
	Exchange Rates	5	6	8	200
	Market situation	9	7	5	315
Staff	Professional skills and abilities	7	8	2	112
	Education	6	5	2	60
	experience	5	6	2	60
Organization of the process	Methods of procurement	9	7	3	189
	Form of procurement organization	9	5	6	270
	Information Support	7	4	4	112
	Methods for evaluating suppliers	8	6	7	336

Table 4: Calculation of the priority risk number (NPR) for factors affecting procurement activity

According to the compiled table 4, it is possible to construct a Pareto diagram reflecting the distribution of individual factors affecting the organization of deliveries on a "just-in-time" system, depending on their significance (Figure 3).

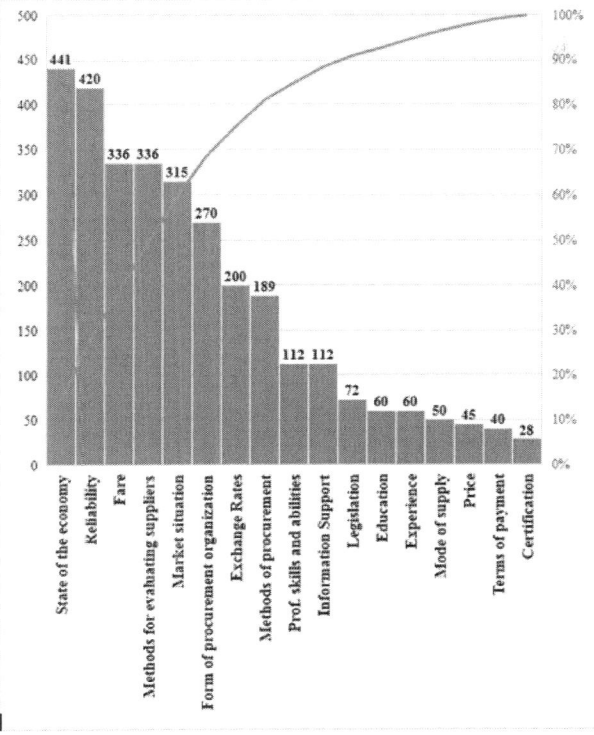

Figure 3: Distribution of individual factors affecting the organization of deliveries on a "just-in-time" system, depending on their significance

Analyzing the distribution diagram of individual factors influencing the organization of deliveries on a "just-in-time" system depending on their significance, built on the basis of calculating the priority number of risks, the following six subfactors can be distinguished:

– state of the economy;
– reliability of suppliers;
– fare;
– methods for evaluating suppliers;
– market situation;
– form of procurement organization.

According to the results obtained, these factors are at greatest risk. The total significance of these six subfactors is 70%, so attention should be paid to reducing the probability of risks arising from them. However, on the state of the economy and the market situation, the possibility of influence is almost minimal, so the key attention should be paid to working with suppliers.

References

[1] Virtual enterprises [Electronic resource]: Logists, site – Access mode: http://www.logists.by/library/view/virtyalnye-predpriyatiya;

[2] Risks in the logistics system [Electronic resource]: Access mode: http://lektsii.org/3-93215.html;

[3] Kazakova, A. I. Optimization of the inventory management system using the "just-in-time" model [Text] / A. I. Kazakova // Economics and management: analysis of trends and development prospects. 2014. № 10 – 80 p.

[4] Life cycle of the product [Electronic resource]: Access mode: http://www.studfiles.ru/preview/2030221/page:6/

APPLYING METHODS OF ARTIFICIAL INTELLIGENCE FOR OPTIMIZATION IN PRODUCTION AND LOGISTICS

M.Sc., M.Sc. Sebastian Lang
Institute of Logistics and Material Handling Systems
Otto von Guericke University Magdeburg, Germany

1   Introduction

In the field of production and logistics, mathematical optimization is an everyday necessity. For instance, in a manufacturing system, which produces a high variety of products, planers have to schedule the production of goods in order to meet customer deadlines. On the other hand, the production plan shall be designed to support a continuous production flow, as well as a high utilization of machines. To give another example for the need of optimization, we can consider a mass production with low variant diversity. In that case, the planning department need to calculate the optimal production volume, which has to be be sufficient to serve future customer demands, but which should also be as small as possible to reduce the number and sizes of stocks and the number of products which are work in process.

The daily optimization problems to solve are often highly complex. Many of those problems are classified as np-hard. The abbreviation *np* stands for nondeterministic polynomial time. This means that only a nondeterministic machine can solve those problems within an acceptable computational time [1]. The reason for this is that the solution space of np-hard problems grows in a non-polynomial manner, while the problem itself is only extended by one variable. In consideration of the fact that a conventional computer can only process data deterministically, the state of scientific knowledge is that no algorithm exists which is able to calculate the provable optimal solution for an np-hard problem without enumerating the complete solution space [2]. Therefore, production planers usually apply specific heuristics or metaheuristics instead to approximate the optimal solution.

However, specific heuristics and metaheuristics have some drawbacks making their application challenging. A specific heuristic is only suitable for a particular problem. Therefore, if planners face a new optimization problem, they need to develop a new heuristic, or they need to investigate whether or not a suitable heuristic already exists. Regardless of whether a new heuristic needs to be developed or if a given heuristic can be

applied, both methods are laborious and require expert knowledge in operations research, programming and the problem itself. Metaheuristics, on the other hand, are theoretically suitable for every kind of optimization problem because they search for solutions without requiring precise information about the problem structure [3]. Still, a problem related adaptation needs a fundamental knowledge about how the metaheuristic works and how adjustments of the metaheuristic's control variables influence the solution process. Furthermore, the non-problem tailored search for solutions can lead to a high computational effort, since every solution candidate need to be evaluated (for instance, by computationally intensive simulation models).

In the current decade, researchers achieved some significant breakthroughs in the field of artificial intelligence [4], [5]. On the one hand, these breakthroughs can be explained by the technical progress of computer hardware, which has a direct influence on the performance and therefore on the potentials for application of AI technologies. On the other hand, AI researchers still invest much effort to improve the existing AI methods. A couple of these methods have some interesting properties, which may make them relevant as an alternative solution approach for optimization problems.

This paper presents the first concept of a solver for optimization problems, further named as AI-Optimization-Framework (AIOF), which combines the AI methods "artificial neural network" and "fuzzy logic" together with discrete event simulation. The upcoming section contains a short summary of the theoretical principles of AI and presents some fundamental AI methods. The third section contains some exemplary related research that utilizes AI methods for optimization problems. The fourth section presents the solver concept, and the last section states future research challenges.

2   Artificial Intelligence

The term *Artificial Intelligence* is not clearly defined, since there are several literature sources providing different explanations of the term [6]. In

[7], the author presents a couple of definitions for artificial intelligence, which arose over time. In his opinion, the following definition is most likely to apply: "Artificial Intelligence is the study of how to make computers do things at which, at the moment, people are better" [8]. This definition seems accurate, because it reflects the current driving trends in AI research. Especially, the further developments of artificial neural networks led to significant improvements of computational image and speech recognition. Those are primary examples, in which humans have always outperformed computers.

In this section, the subject of interest is how to make computers "intelligent". The field of AI provides several methods for this purpose. Similar to the definition of AI, the research community does not agree about how to classify AI methods. Therefore, one can find various methods in the literature which claimed to be AI, but which in turn other researchers do not consider as such. In this paper, to classify AI methods, we rely on the collective name *Computational Intelligence* (CI), which "[...] comprises concepts, paradigms, algorithms, and implementations to develop systems that exhibit intelligent behavior in complex environments" [9]. CI includes three fundamental methods [10]:

− Artificial Neural Networks (ANN)
− Fuzzy Logic
− Evolutionary Algorithms (EA)

According to [11], an ANN is a "[...] massively parallel distributed processor made up of simple processing units, which has a natural propensity for storing experiential knowledge and making it available for use". A processor unit is called a *Neuron*. Neurons can process multiple input streams and output one result. To accomplish this, the activation function within the neuron calculates the output value, considering the given input data. Neurons are organized in so-called layers. An ANN consists of at least one layer. However, more usual are network structures with three or more layers, whereby one input layer directly processes incoming data and one output layer represents the space of possible solutions. A neuron maintains connections to other neurons, which usually belong to adjacent layers. These weighted connections are called *Synapses*. They determine how much a neuron reacts on value changes of an upstream neuron. ANN apply machine-learning algorithms to find a relationship between input and output data. For this purpose, the majority of machine-learning algorithms analyze several reference data sets of a problem and adjust the synapse weights of the ANN iteratively, until no significant improvement is possible anymore. After successfully completing a sufficiently large training period, an ANN is able to estimate output values for similar problems.

Fuzzy logic is an alternative approach to conventional boolean logic. It provides a formalization of *Approximate Reasoning* and is closely connected to the theory of fuzzy sets, which describe sets of related classes with indistinct boundaries [12]. The principles of fuzzy logic allows a machine to transform qualitative information into quantitative data and vice versa. The main idea of fuzzy sets is to create a link between linguistic and numeric variables [13]. In this context, each value of a linguistic variable corresponds to an interval of values of the numeric variable. In contrast to conventional quantization of values, the boundaries of the intervals within a fuzzy set are overlapping. As an example, let us introduce the linguistic variable *Temperature*. We assume that *Temperature* can obtain the values "cold", "warm" and "hot". The corresponding numerical intervals are [0, 35], [25, 60] and [50, 100] centigrade. As we can see, temperatures between 25 and 35 centigrade can be considered as "cold" or "warm" and temperatures between 50 and 60 centigrade as "warm" or "hot". For the resulting fuzzy intervals [25, 30] and [50, 60], there are two probability functions describing the most likely membership of each possible numeric value to each possible linguistic value. Thinking of a software implementation, an algorithm can use this information to determine randomly the real membership of the numeric value. Describing linguistic variables with fuzzy sets rather than strictly defined intervals seems more reasonable, since fuzzy sets consider the subjectivity of perception.

The term *Evolutionary Algorithm* describes a class of optimization methods, which search for solutions according to the behavior of biological evolution [14]. Popular examples for EA are genetic algorithms, scatter search and ant colony optimization, which are detailed in [15]. Each EA has its own scheme to search for an optimum, but the basic process is always the same.
In an initial step, the EA randomly creates a set of start solutions also known as population. Despite the randomness of the process, the EA tries to meet specific requirements for the initial population, for instance maintaining a large-scale distribution of population members across the complete solution space. After the initialization, the EA enters the iterative search process. In the first step, the EA generates a subset of solutions with the best results. Based on the members of this subset, the EA creates in the second step new solution candidates. For this purpose, the EA can "mutate" existing solutions, which means that the solution vector of the candidate changes in some specific manner. Furthermore, the EA can "pair" two members of the subset, which means that the algorithm combines the vectors of the "parent" solutions in a certain way to generate a "child" solution. The EA saves each mutated or new generated solution candidate within the

subset. In the third step, the EA replaces a specific number of population members, which provide the worst solutions, with a corresponding number of subset members providing the best solutions. Afterwards, the EA deletes the subset and may goes into the next iteration.

3   Related Research

As shown in the previous section, the field of AI methods is broad and a comprehensive literature review would go beyond the scope of this paper. This section only provides a highly aggregated overview of related publications. The investigation results are part of a systematical literature review, which is still in process. Therefore, the papers referenced in this section do not represent a holistic view on the research about "AI for optimization", but rather a first insight. Furthermore, this section will not discuss publications describing the application of EA for optimization. As described in the introduction to this paper, the application of metaheuristics is considered to be state of the art for solving optimization problems. Since every EA can be considered as metaheuristic [16], a literature review about EA for optimization would be pointless. A comprehensive review about evolutionary algorithms and other metaheuristics for combinatorial optimization problems can be found in [17].

There are several papers presenting the application of ANN for optimization in production and logistics. However, almost every publication deals with a scheduling problem, by which the proposed solution approach shall determine an optimal assignment of customer orders to machines or an optimal sequence of already assigned customer orders. So far, three different approaches could be identified to apply ANN:

1. A set of priority dispatching rules controls the operations in the production system. Only one priority dispatching rule can be enabled at the same time. In this scenario, an ANN selects for a specific period of time an appropriate dispatching rule by analyzing the current system state. For instance, [18], [19] and [20] presents this approach in their work.
2. The authors apply ANN to estimate parameters for a self-developed priority dispatching rule. The estimated parameters are coefficients of the priority index function, which have an unknown relation to the problem itself [21].
3. Another possibility to utilize ANN for scheduling problems describes [22]. Here, the ANN decides directly on which machine a customer order should be allocated. The decision is based on attributes of the customer order, such as due date or product family, as well as, the current system state.

Until now, only three papers could be investigated, which describe fuzzy-logic based approaches for optimization or decision-making. Reference [23] presents a fuzzy-logic based expert system to evaluate solutions of a simulation optimization framework. The authors argue that determining the precise performance limit of a system, for having a benchmark to evaluate solution candidates, is for complex problems not feasible. Instead, expert knowledge of analysts and decision makers, represented by linguistic variables, can also express the quality of a solution. In [24], the author applies fuzzy logic for labor allocation within a manufacturing system. Here, the queue size in front of a machine and the intensity of the corresponding process are the subjects of two membership functions. By calculating the product of the outputs from both functions, the simulation model receives the probability for assigning a worker to the corresponding machine. The simulation model decides to allocate a worker, if a randomly generated number is smaller than the calculated probability. The authors of [25] propose a fuzzy-logic based inventory control system to consider the dynamic, stochastic and uncertainty of the customer demand. The task of the inventory control system is to decide about the production amount in order to maintain a desired inventory level, which varies over time. The system performs the decision depending on the difference between the current desired and the actual inventory level. The authors apply fuzzy logic to assess the size of this difference in comparison with tracked differences from the past. Based on the identified membership, a specific correction parameter adjusts the desired production amount. In conclusion, the proposed system analyses desired inventory levels from the past, in order to determine a production amount, which will presumably meet future fluctuations of the desired inventory level.

4   A First Concept for an AI-based Optimization Framework

Compared to metaheuristics, the solution approaches presented in the previous section suffer from some shortcomings. Looking on the ANN-based solution approaches, one may criticize that the rule-based allocation and sequencing of customer orders does not respect the characteristics and particularities of a specific optimization problem. Metaheuristics do not consider both as well, but instead they are able to search the whole solution space of an optimization problem for promising solution candidates. Another drawback of the ANN-based solution approaches is the merely partial view on the optimization problem itself during the decision-making process. More precisely, the described approaches are considered as real-time scheduler, which only analyze the current system state and the attributes of the customer order to be assigned, when a decision needs to be processed. Therefore, those approaches do not allow a holistic view on an optimization problem, i.e. they are unable to identify potential synergies and influences of a performed decision on upcoming customer orders.

Hence, the AIOF has to fulfill the following requirements:

1. Like metaheuristics, the AIOF should suggest completely described solutions for a given optimization problem with a given input. This excludes, for instance, the recommendation of a system behavior described by a priority dispatching rule.
2. During the calculation of a solution, the AIOF has not only to consider the current state of the system and the attributes of the currently analyzed decision object. For a local decision, it should also be able to consider important information about decision objects, which it would analyze at a later point of time. Furthermore, the framework has to assess the value of a local decision, before settling that decision, for instance by estimating the resulting system state.
3. The computational time for calculating a result has to be sufficiently low to face operational decisions with a short-term due date. Since there is no general definition of when a due date is considered as short-term, the user of the AIOF needs to be able to scale the investment of computational effort and computational time for calculating a result.

Metaheuristics are also scalable in terms of computational effort and computational time. For instance, metaheuristics allow to determine a maximum time limit to search for solutions or a maximum number of solution candidates to evaluate. However, the probability for finding the optimal solution falls with the invested time to search for solutions. In contrast, the solution quality of ANN and fuzzy logic is not related to the available computational time, because both methods determine a solution by experiential estimating instead of searching. From this point of view, a fourth requirement results that concerns the significance and reasonability of the AIOF:

4. If only a minor time span is available to setup, and carry out an optimization process, the AIOF has to outperform any searching metaheuristic in terms of solution quality.

The concept and design of the proposed framework shall reflect these requirements. Figure 1 presents the basic structure of the AIOF.

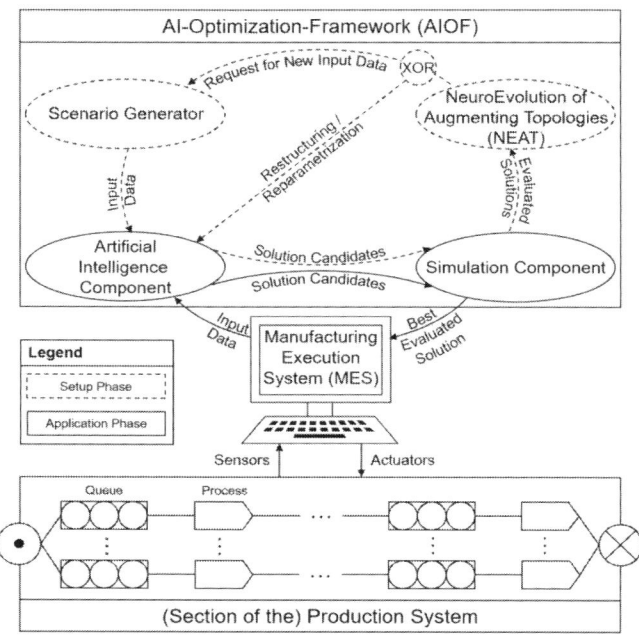

Figure 1: Conceptual Model of the AI-Optimization-Framework

As illustrated, the framework consists of four components:

- An AI component for optimization mainly based on an ensemble of ANN. The ANN differ in terms of their network structure and synapse weights.
- A simulation component for evaluating the Solution candidates
- A machine-learning component based on *NeuroEvolution of Augmenting Topologies* (NEAT). NEAT relies on an adapted genetic algorithm, which simultaneously search for optimal ANN network structures and optimal synapse weights. A detailed description about the NEAT approach provides [26].
- A scenario generator to create training data, based on the production system to be analyzed and the optimization problem to be solved

Before a production system can apply the AIOF, the AI component requires a *Setup Phase*. Within the setup phase, the AI component receives training data from the scenario generator. In the next step, every ANN of the ensemble suggests a problem solution, based on its specific structure and synapse weights. The simulation component sequentially utilizes the corresponding output data to calculate user-specified KPIs. Furthermore, the results of each simulation experiment are aggregated to a general system performance indicator. Afterwards, the NEAT method checks, if the system performance indicator exceeds a specific threshold. If false, NEAT adjusts the network structure and synapse weights of the corresponding ANN. If true, NEAT sends a message to the scenario generator to create a new problem instance. Once the NEAT method has adjusted the AI component for a satisfying number of problem instances, the AIOF is prepared for the *Application Phase*. Within the application phase, the AIOF receives input data from the MES of the production system. Still, the simulation component evaluates the quality of each solution candidate generated by a specific ANN of the ensemble. The AIOF will send back the most promising solution candidate as result to the MES.

A successful application of the AIOF mainly depends on the AI component, which represents the core element of the whole framework. So far, there are several ideas about the design of the AI component. However, this paper will only contain the description of the most promising concept so far. For this purpose, we want to consider an order-sequencing problem with a planning horizon of 40 customer orders. The input layer of each ANN receives the attributes of one customer order at the same time. Furthermore, the input layer contains a subset of neurons to consider the current system state. The number of neurons in the output layer represents the solution space of the optimization problem. In our example, the output layer of each ANN consists of 40 neurons, whereby each neuron represents one order position. The determination of a solution is an iterative process. In the described scenario, the ANN requires 40 iterations to determine the complete solution vector. In each iteration, the ANN estimates the specific position of a customer order. During the solution process, it may happens that the ANN tries to allocate a customer order to an already assigned position. To avoid this problem, the output layer of an ANN utilizes fuzzy logic to express the certainty of each allocation. If a conflict situation occurs, an algorithm compares the allocation certainty of the two affected customer orders. The algorithm assigns the customer order with the highest allocation certainty to the conflicted order position. The remaining customer order will be allocated to the next most suitable order position.

5   Future Research Challenges

So far, there are still some open problems, which the current concept ideas cannot yet resolve:

– What can be implemented so that the AIOF considers mathematical constraints of the optimization problem during the solution process? Since a trained ANN is only able to estimate exactly one solution for a specific input data stream, it is highly important that the resulting solution is feasible.
– How should the AIOF react on unexpected changes regarding the statistical properties of input data? Significant changes of input data may lead to unsuitable solutions. Therefore, it is important to develop an adaption strategy, which is able to quickly adjust the parameters of an ANN

– Which KPIs are meaningful to represent the attributes of upcoming decision objects, while the AIOF determines a decision for a currently treated decision object? The consideration of such KPIs as input neurons could be important to maintain a holistic view on the optimization problem and to perform foresighted decisions.

These research items as well as a prototypical implementation of the AIOF will be subject of future publications.

6   References

[1] Zimand, M. (2004): Computational Complexity: A Quantitative Perspective. Amsterdam: Elsevier B.V., p. 52.
[2] Hromkovič, J. (2014): Theoretische Informatik. Formale Sprachen, Berechenbarkeit, Komplexitätstheorie, Algorithmik, Kommunikation und Kryptographie. Wiesbaden: Springer Vieweg, p. 190.
[3] Blum, C.; Roli, A. (2003): Metaheuristics in Combinatorial Optimization: Overview and Conceptual Comparison. ACM Computing Surveys 35 (3): 268–308.
[4] Krizhevsky, A.; Sutskever, I.; Hinton, G. E. (2012): ImageNet Classification with Deep Convolutional Neural Networks. In: NIPS'12 Proceedings of the 25th International Conference on Neural Information Processing Systems - Volume 1. Lake Tahoe, NV, USA, pp. 1097–1105.
[5] Silver, D.; Huang, A.; Maddison, C. J.; Guez, A.; Sifre, L.; van den Driessche, G. et al. (2016): Mastering the game of Go with deep neural networks and tree search. Nature 529 (7587): 484–489.
[6] Jang, J.-S. R.; Sun, C.-T.; Mizutani, E. (1997): Neuro-Fuzzy and Soft Computing: A Computational Approach to Learning and Machine Intelligence. Upper Saddle River: Prentice Hall, pp. 3–5.
[7] Ertel, W. (2016): Grundkurs Künstliche Intelligenz: Eine praxisorientierte Einführung. Wiesbaden: Springer Vieweg, p. 2.
[8] Rich, E. (1983): Artificial Intelligence. New York: McGraw-Hill
[9] Kruse, R.; Borgelt, C.; Braune, C.; Mostaghim, S.; Steinbrecher, M. (2016): Computational Intelligence: A Methodological Introduction. London: Springer, p. 2.
[10] VDI/VDE 3550 Blatt 1 / Part 1 (2001): Computational Intelligence: Künstliche Neuronale Netze in der Automatisierungstechnik: Begriffe und Definitionen / Artificial neuronal network in automation: Terms and definitions. Berlin: Beuth, p. 2.

[11] Haykin, S. (1999): Neural Networks: A Comprehensive Foundation. Dehli: Pearson Education, pp. 23, 32 ff.

[12] Zadeh, L. A. (1994): Fuzzy Logic, Neural Networks, and Soft Computing. Communications of the ACM 37 (3): 77–84.

[13] Zadeh, L. A. (1975): The Concept of a Linguistic Variable and its Application to Approximate Reasoning–II. Information Sciences 8 (4): 301–357.

[14] Weicker, K. (2015): Evolutionäre Algorithmen. Wiesbaden, DE: Springer Vieweg, p. 2.

[15] Gendreau, M., Potvin, J.-Y. (2010): Handbook of Metaheuristics. New York: Springer Science+Business Media.

[16] Burke, E. K.; Kendall, G. (2005): Search Methodologies. Introductory Tutorials in Optimization and Decision Support Techniques. New York: Springer Science+Business Media, p. 16.

[17] Blum, C.; Roli, A. (2003): Metaheuristics in Combinatorial Optimization: Overview and Conceptual Comparison. In: ACM Computing Surveys 35 (3): 268–308.

[18] Arzi, Y.; Iaroslavitz, L. (1999): Neural Network-Based Adaptive Production Control System for a Flexible Manufacturing Cell under a Random Environment. In: IIE Transactions 31 (3): 217–230.

[19] Azadeh, A.; Maleki Shoja, B.; Moghaddam, M.; Asadzadeh, S. M.; Akbari, A. (2013): A Neural Network Meta-Model for Identification of Optimal Combination of Priority Dispatching Rules and Makespan in a Deterministic Job Shop Scheduling Problem. In: The International Journal of Advanced Manufacturing Technology 67 (5–8): 1549–1561.

[20] Bergmann, S.; Stelzer, S.; Strassburger, S. (2014): On the Use of Artificial Neural Networks in Simulation-Based Manufacturing Control. In: Journal of Simulation 8 (1): 76–90.

[21] Park, Y.; Sooyoung, K.; Young-Hoon, L. (2000): Scheduling Jobs on Parallel Machines Applying Neural Network and Heuristic Rules. In: Computers & Industrial Engineering 38 (1): 189–202.

[22] Hammami, Z.; Mouelhi, W.; Lamjed, B. S. (2017): On-line Self-Adaptive Framework for Tailoring a Neural-Agent Learning Model Addressing Dynamic Real-Time Scheduling Problems. In: Journal of Manufacturing Systems 45: 97–108.

[23] Medaglia, A. L.; Fang, S.-C.; Nuttle, H. L. W. (2002): Fuzzy Controlled Simulation Optimization. In: Fuzzy Sets and Systems 127 (1): 65–84

[24] Pugh, G. A. (1997): Fuzzy Allocation of Manufacturing Resources. In: Computers & Industrial Engineering 33 (1–2): 101–104

[25] Samanta, B.; Al-Araimi S. A. (2001): An Inventory Control Model Using Fuzzy Logic. In: International Journal of Production Economics 73 (3): 217–226

[26] Stanley, K. O., Miikkulainen, R. (2002): Evolving Neural Networks through Augmenting Topologies. In: Evolutionary Computation 10 (2): 99–127

METHODOLOGY FOR THE MANAGEMENT OF RISK IN THE STORAGE AND TRANSPORT OF HAZARDOUS SUBSTANCES

Ing. Lissette Concepción Maure
Industrial Engineering Department, Central University from Las Villas, Cuba

DrC. Félix Abel Goya Valdivia
Chemical Engineering Department, Central University from Las Villas, Cuba

Prof. Dr.-Ing. Dr. h.c. Norge Isaias Coello Machado
Mechanical Engineering Department, Central University from Las Villas, Cuba

Dr.-Ing. Dr. h.c. (UCLV) Elke Glistau
Institute of Logistics and Material Handling Systems
Otto von Guericke University Magdeburg, Germany

Abstract

The decision making has great importance in the formulation of prevention and recovery policies against technological accidents in the chemical process industry and companies that handle hazardous substances. The main objective of management of technological risks in storage and transport activities along the supply chain, is the search of alternatives to reduce or mitigate the major hazards without eliminating the obtention of benefits. The objective of this research is to develop a general procedure and its methodological instruments for the management of risks of major accidents in activities of storage and distribution of hazardous substances. It includes multicriteria analysis, risk measurement methods and control tools to identify, characterize and hierarchize the storage areas and distribution routes of greater danger. The application of the procedure enables the reorientation of organizational efforts supported by information technologies and ensures a continuous improvement approach. This research takes as case of practical study the logistics network of the Fuel Trading Company of Villa Clara and uses the strategy of multiple explanatory cases in different companies that operate with hazardous substances in the province.
Keywords: risk management; storage and transport of hazardous substances; multicriteria analysis; control tools

1   Introduction

Modern industry is characterized by continuous growth of the unitary power on its plants to obtain better performance[6]. Regardless of the scientific technical development, the increase in the complexity's degree of technological processes generates risk conditions in society and natural environment that acts as support for it [1, 7]. Given this reality, the paradigm of technological risk management and the conceptual approach (social, economic and environmental) that underlies it, have evolved from the theoretical point-of-view in a remarkable way [10].

The importance of risks management in the handling of hazardous substances is given by the following aspects: production increase on products of high added value, which require industrialized processes with narrow safety margins [11]; increase of inventories [5]; diversity of distribution routes, change in risk profiles of the supply chain as a result of changes in their business models [9]; population growth that leads to an unplanned urbanization near the industrial sector [12]; the inclusion in the organizational performance of the sustainable development concept [3]; the need to ensure the efficient and optimal allocation of limited resources in processes of evaluation and risk management [12].

On the literature review, the research problem was defined as the lack of a prescriptive theory for analysis of major hazards in logistic activities of storage, processing and distribution of dangerous substances.

The decision making in the logistic processes when dangerous substances are handled requires that the risk is measured and represented by models, maps and indices. These should consider the existing dangers, the vulnerability of the system, the expected physical damage and the possible aggravation of the impact according to social, economic and environmental conditions.

Reference [2, 13] consider that the main objective of the management of technological risks within the logistics process is the search of alternatives to reduce or mitigate the major hazards without eliminating the obtention of benefits. In this regard, a multicriteria analysis is necessary to manage the uncertainty regarding a threat and the vulnerability of the system. This must be done through a sequence of activities that include the identification of triggering events, prevention and mitigation actions, levels of acceptability, disaster management, governance and transfer.

Despite the importance given by the government, the academic circles and the business sector, Cuba recognizes the lack of a framework that analyzes the complexity of the major technological risks in the supply chains. The absence of a holistic conception and a systemic and continuous improvement approach, which addresses all dimensions of risk management, limits a modification of the situation reflected.

The present research shows a procedure for the management of technological risks in activities of storage and distribution of hazardous substances as support for the decision making process. This document is structured in five sections. In the next section, a research background of the models and indices of evaluation of technological risk, and a discussion about its advantages and limitations is presented. Section 3 proposes a methodology for management of technological risks in logistics processes where hazardous substances are handled. Results and conclusions are presented in section 4 and 5.

2   Research background

Several methodologies have been developed to study technological risks in logistics processes. According to different probable risk scenarios and their interaction with the environment, those methodologies have progressed towards a dynamic direction [8].

Technological risk management depends on the measurement of the level of risk associated with the identified hazards. It also depends on the degree of precision with which the variables that condition it and its synergy are determined [11]. The risk profiles should show the existing situation and allow the classification and prioritization of activities.

The choice of risk metrics is critical since it selects the type of information included in the study and legitimizes the results [9]. Consequently, the assessment of risk level must be deployed by various levels of analysis: risk activities, logistics processes and supply chains. Some advantages reported in the reference [3, 4, 11, 13] of use of risk indexes in security management systems are:

–   Reducing the complexity of risk management at the company level and make it possible to measure their social and environmental performance. The information is synthesized and expressed by a numerical value including parameters and/or variables of risk management.

–   Evaluate and support decisions regarding environmental and social impact allowing the observation of evolution in the time and study trends about disaster situation.

–   Fulfillment of accomplish with social and environmental laws.

–   Operability of the strategies. It shows the limits for acceptable operations that can lead to better efficiency of process and serve as basis for planning inspections and establishing prevention measures.

–   Improvement of performance. It facilitates internal communication and helps to mantain a high degree of awareness about prevention of major accidents. Facilitates the efficient and optimal allocation of limited resources for risk assessment around the classification and prioritization of different scenarios.

Reference [2] states that determining the level of risk requires the use of different mathematical and empirical models. Reference [8] provides an explanatory overview of risk metrics related to the study of major accidents. At the same time, it shows in most cases these are conditioned to estimate certain variable within the risk assessment process, making it difficult to prioritize the sources of technological risks within a supply chain.

3   Methodology to determine the level of technological risk in logistics processes

In this section we will show how to determine the current risk level in the logistics processes. To achieve this goal we will follow the procedure shown in figure 1.

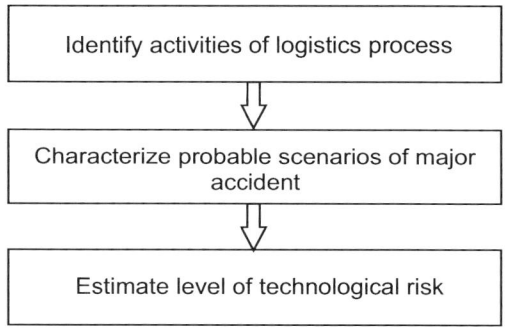

Figure 1: Methodology to determine the level of technological risk

3.1   Identify activities of logistics process

This step constitutes the basis to determinate the scenarios of major accident occurrence considering the hazard of technological risk. All activities and relations between different organizations belonging to the supply chain are delimited. Once the flows of existing materials have been analyzed, a unique inventory of hazardous substances is made. This inventory relates all substances with potential to trigger a major accident, causing damage to people (workers and surrounding communities), industrial or public property and environmental components. The experts will assess the physical-chemical nature of inventoried substances and type of potential damage (explosives, flammable-toxic liquids, flammable-toxic gases), forms of containment, associated activities (storage, processing, distribution), possible initiating events, disasters events that can be triggered and routes of propagation.

3.2 Characterize probable scenarios of major accident

In this step, the group of experts must establish a sequence of accidents that can be triggered considering the occurrence of an initiating event:

- Spillage of toxic liquids: due to loss of fluid containment, it can generate toxic effects, fires and/or explosions, depending on the nature of the substances.
- Exhaust of gases: due to loss of fluid containment, it can generate toxic effects, fires and/or explosions, depending on the nature of the substances.
- Fire: combustion of multiple forms of the contained or emitted fluids generates harmful thermal radiation, when the substances are flammable.
- Explosion: prior to the emission or after the fire, generates pressure or overpressure waves, and the propagation of projectiles.

The process is supported by the software ALOHA (Areal Locations of Hazardous Atmospheres). This computer program designed for models key hazards-toxicity, flammability, thermal radiation (heat), and overpressure (explosion blast force) - related to chemical releases that result in toxic gas dispersions, fires, and/or explosions. Its chemical library contains information about the physical properties of approximately 1 000 common hazardous chemicals.

ALOHA allows to determinate the radius of affectation in the event of a major accident taking into account: type of substance, form of containment and description of how the chemical is escaping from containment, and weather conditions. The software will display the threat zones in red, orange, and yellow. The red threat zone represents the worst hazard and the orange and yellow threat zones represent areas of decreasing hazard.

3.3 Estimate level of technological risk

In this step, the level of technological risk will be assessed in logistics activities. The risk level estimation must quantify the damage caused within the affected radius, delimited in the previous step. A holistic assessment of risk takes into account: 1) the physical damage: number of victims and economic and environmental losses (first-order effects) 2) the conditions related to the social fragility and the resilience lack of communities that favor the occurrence of accident or aggravate the impact of these (second-order effects).

The analytical structure of indicators systems for holistic evaluation of technological risk (IRT) in an

activity i is expressed as the sum for each possible event e (fire, explosion, spill, escape), considering their occurrence probability p_e and probable physical consequences C_e within the radius of affectation. It is affected by a coefficient of aggravation of the impact Cai, which depends on conditions of socioeconomic fragility and lack of resilience of the community (equation 1).

$$IRT_i = (1 + Cai_i) \sum(p_{ei} * C_{ei}) \tag{1}$$

The consequences respond to the determination of the physical damage before an event e in activity i. This is evaluated using the equation 2.

$$C_{ei} = \sum_{n=1}^{p} w_{X C_{ne}} * X_{C_{ne}} \tag{2}$$

Where $X_{C_{ne}}$ represents the physical risk factors, $w_{X C_{ne}}$ the weights of these factors and p is the total number of factors to be considered in the calculation. We propose the quantification of victim's number, economic losses and environmental damage, with an equivalent weight. The coefficient of aggravation Cai_i depends on the weighted sum of a set of aggravating factors in the social, economic, ecological, structural, nonstructural and functional perspective; associated with the fragility of community X_{FSi} and the resilience lack of context X_{FRj}, being w_{XFSi} and w_{XFRj} the weights of each factors.

$$Cai = \sum_{i=1}^{m}(w_{XFSi} * X_{FSi}) + \sum_{j=1}^{n}(w_{XFRj} * X_{FRj}) \tag{3}$$

The evaluation results of analysis units are presented in terms of relative indexes of physical risk, socioeconomic fragility, resilience lack of and total risk.

The set of descriptors used in the multicriteria evaluation corresponds to qualitative or quantitative data that are derived from previous studies, damage scenarios and socio-economic information of the context to be analyzed.

The descriptors proposal was made based on a bibliographic compilation of risk indicators proposed by different methodologies to assess physical risk, socioeconomic fragilities and resilience lack (table 1).

Perspective	Descriptors	Unit	Criteria
Social	Population density	People/ha	X_{FS}
	Presence of community areas	%	X_{FS}
	Level of human development	Index (0,1)	X_{FR}
	Reaction capacity	Index (0,1)	X_{FR}
	Perception of risk	Qualification (1,2,3,4,5)	X_{FS}
Ecological	Vulnerable environmental receptors	%	X_{FS}
	Reversibility of damage - recovery	%	X_{FR}
Economic	Potential losses	$	X_{FS}
	Financial resilience	$	X_{FR}
	Institutions within the possible radius of affectation	%	X_{FS}
Structural	Physical condition of constructions	Qualification (1,2,3,4,5)	X_{FS}
	Nearby facilities that handle hazardous substances	%	X_{FS}
	Protection of facilities	Qualification (1,2,3,4,5)	X_{FS}
	Evacuation system	Qualification (1,2,3,4,5)	X_{FR}
	Structural reconstruction	year	X_{FR}
Not structural	Presence of aggravating non-structural units	%	X_{FS}
	High density traffic routes	%	X_{FS}
	Non-structural reconstruction	year	X_{FR}
Functional	Security practice	Qualification (1,2,3,4,5)	X_{FR}
	Emergency plans (internal and external)	Qualification (1,2,3,4,5)	X_{FR}
	Operability of the emergency	Minutes	X_{FR}
	Firefighting brigades	1/1.000 hab	X_{FR}
	Hospital services	1/1.000 hab	X_{FR}

Table 2: Descriptors of socioeconomic fragility and lack of resilience

These descriptors used in holistic risk assessment have different units. To standardize the gross value of the descriptors, transforming them into commensurable values, must be used transformation functions with the pattern shown is Figure 2.

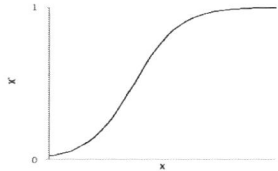

Figure 2: Sigmoidal type transformation function for the normalization of risk indicators

The previous function responds to the equation 4.

$$X' = \frac{1}{1 + e^{-\beta\left(\frac{X-m}{M-m} - \mu\right)}} \qquad (4)$$

Where,
X: Initial value of the indicator
X': Normalized value of the indicator
e: Base of natural logarithm
β: Parameter - slope of the curve
M: Maximum value of parameters (descriptors from table 1)
m: Minimum value of parameters (descriptors from table 1)
μ: Point of inflection of the curve

The parameters valuesused for the transformation of each descriptors are obtained from the reference values established by experts, bibliographic review, observations made in major accidents and examination of descriptor statistics along the chain.

The weights of the descriptors represent the relationships of hierarchy (relative importance) in the aggregation process through a multicriteria evaluation. The evaluation of these coefficients is carried out through the analytical hierarchical process (AHP). This is based on the comparison between pairs of descriptors to establish the relative importance (quantitatively). These comparisons generate a matrix that allows to calculate the weight factors and verify the exercise consistency. As a result, we obtain a set of weight factors that are less sensitive to judgment errors.

A network can be generated from a hierarchy by gradually increasing the interconnections. This allows to generate a network, taking into account all existing relationships between levels (perspectives) and between alternatives (descriptors) without assuming the axiom of dependence. At the same time, it generates maps of causal relationships, with a solid mathematical foundation. The figure 3 shows the analytical network modeled in the SuperDecisions software.

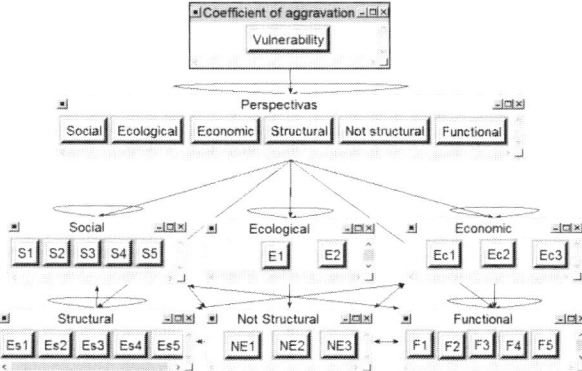

Figure 3: Weight`s network of socioeconomic fragility descriptors and resilience lack descriptors

Table 2 presents the results of application of AHP method.

Perspective	Weighting coefficient from the perspective	Code	Weighting coefficient	Equivalent weighting coefficient Wi Wi
Social	0,240	S1	0,326	0,078
		S2	0,246	0,059
		S3	0,108	0,026
		S4	0,160	0,038
		S5	0,160	0,038
		Σ	1,00	
Ecological	0,124	E1	0,660	0,082
		E2	0,340	0,042
		Σ	1,00	
Economic	0,198	Ec1	0,493	0,098
		Ec2	0,196	0,039
		Ec3	0,311	0,062
		Σ	1,00	
Structural	0,144	Es1	0,215	0,031
		Es2	0,140	0,020
		Es3	0,287	0,041
		Es4	0,252	0,036
		Es5	0,106	0,015
		Σ	1,00	
Not structural	0,078	NE1	0,493	0,039
		NE2	0,311	0,024
		NE3	0,196	0,015
		Σ	1,00	
Functional	0,216	F1	0,326	0,070
		F2	0,143	0,031
		F3	0,212	0,046
		F4	0,108	0,023
		F5	0,212	0,046
		Σ	1,00	
Σ	1,00			1,00

Table 2: Weight`s coefficients network of socioeconomic fragility descriptors and resilience lack descriptors

4   Results

In this section the results will be shown according the methodology established in the previous section. This model uses the strategy of multiple explanatory cases in different companies that operate with hazardous substances in the province of Villa Clara. The provincial is subdivided into 13 municipalities, with a total of 124 evaluated facilities.

The inventory of hazardous substances in the province and the evaluation of the activities carried out (storage, processing and distribution) allowed the analysis of 240 potential hazards. This inventory is shown in Table 3.

Cities	Escape of gases toxic	% Total	Fire	% Total	Explosion	% Total	Spillage	% Total	Total
Corralillo	2	14.3	6	42.9	6	42.9	0	0	14
Quemado de Güines	0	0	3	50.0	3	50.0	0	0	6
Sagua La Grande	4	14.8	8	29.6	8	29.6	7	26	27
Encrucijada	1	11.1	4	44.4	4	44.4	0	0	9
Camajuaní	2	25.0	3	37.5	3	37.5	0	0	8
Caibarién	4	26.7	5	33.3	5	33.3	1	6.7	15
Remedios	4	26.7	5	33.3	5	33.3	1	6.7	15
Placetas	3	21.4	5	35.7	5	35.7	1	7.1	14
Santa Clara	7	8.0	39	44.8	39	44.8	2	2.3	87
Cifuentes	1	20.0	2	40.0	2	40.0	0	0	5
Santo Domingo	4	16.7	9	37.5	9	37.5	2	8.3	24
Ranchuelo	1	14.3	3	42.9	3	42.9	0	0	7
Manicaragua	1	11.1	4	44.4	4	44.4	0	0	9
Villa Clara	34		96		96		14		240
%	14		40		40		6		

Table 3: Probable scenarios of major accident in Villa Clara [4]

The total risk is evaluated in each analysis units as a function of exposure factor, (social, economic and environmental consequences) and the aggravating factor through Equation 1.

The figure 4 shows the results of the evaluation carried out in companies located in Villa Clara, divided by municipalities. In this the possible radio of affectation is delimited, and the evaluation of the level of risk is expressed in low, medium and high scale (green, yellow and red). When comparing the results of technological risk in four possible scenarios of major accident, it is observed that Santa Clara municipalities has the highest technological risk index. On the other hand, the municipalities of Quemado de Güines, Camajuaní and Ranchuelo are those exposed to a lower level of technological risk. The figure 4-7 shows the affectation radio of different possible accident.

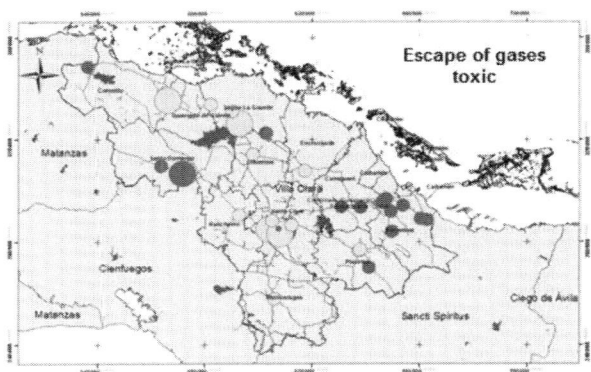

Figure 4: Radio of affectation and level risk. Exhaust of toxic gases[4]

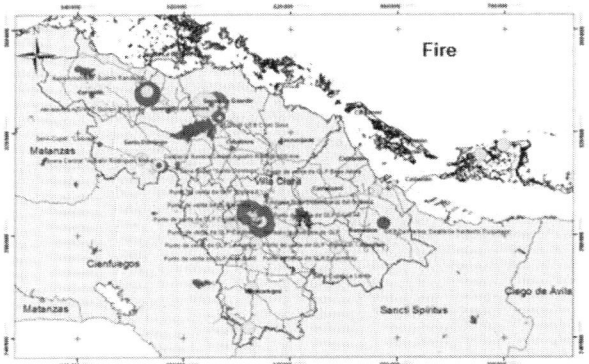

Figure 5: Radio of affectation and level risk. Fire [4]

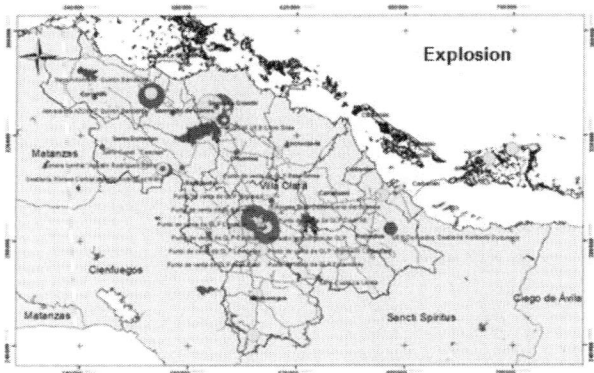

Figure 6: Radio of affectation and level risk. Explosion [4]

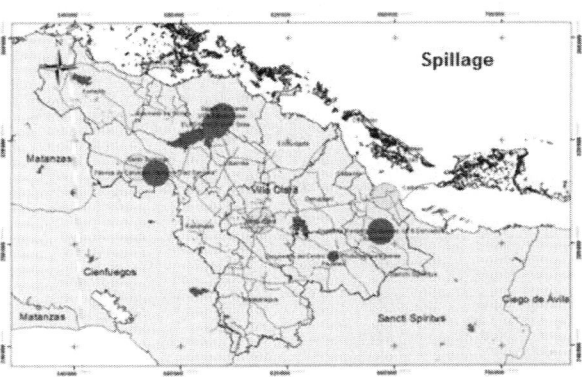

Figure 7: Radio of affectation and level risk. Spillage [4]

This research takes as case study the logistics network in Fuel Trading Company of Villa Clara. This logistics network includes the Fuel Trading Company and the technological warehouse of liquefied petroleum gas (LPG), 53 gas station, 11 stores the sell gas (LPG). These analyzed entities constitute fuel storage and sale centers. This logistics network includes a total of 22 routes by highways and 2 routes by railways.

The highest risk index in storage activities is in the storage area from the Fuel Trading Company, and the most dangerous route is the RFC-02 route corresponding to the transportation of fuel by trains from the Camilo Cienfuegos Refinery in Cienfuegos, to warehouse of the Fuel Trading Company in Santa Clara. This route crosses the center of the town of Cruces, which increases the index of associated vulnerability factor. These results are shown in figure 8 and 9 of technological risk level.

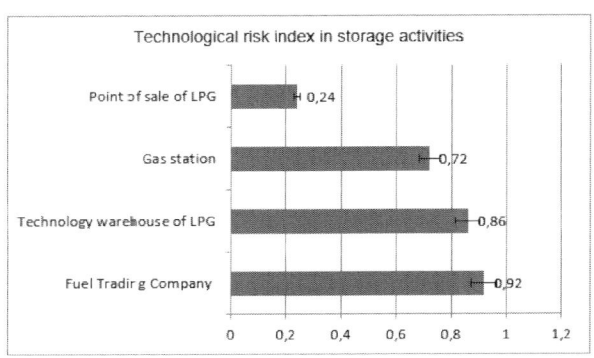

Figure 8: Technological risk index in storage activities [4]

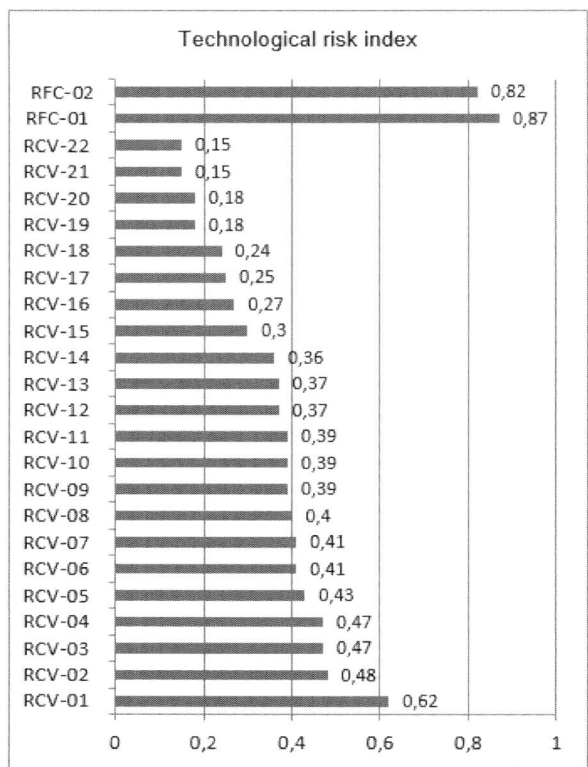

Figure 9: Technological risk index by distribution route [4]

This analysis allows us to index those logistic activities that constitute a major danger in their execution, being necessary to establish disaster prevention and mitigation measures.

Conclusions

The proposed technological risk index considers the effect of existing physical risk given the occurrence of a destabilizing event, as well as the worsening of the impact due to socioeconomic conditions and the resilience lack of the involved area. It provides a scientific basis for risk-based approach and the development of a proactive culture of prevention, improvement and protection.

The indexing of the technological areas and plants depends on existing risk level in the occurrence of major technological accidents. At same time, facilitates the documentation of involved processes in risk management and decision making for the planning of preventive actions.

The analysis of technological risk level in storage and transport activities supports the decision making process. This analysis is based on the characterization and hierarchization of storage areas and distribution routes of greater danger. The application of the procedure allows the reorientation of the organizational efforts and guarantees an approach of continuous improvement.

5   References

[1] Abrahamsen, E.; Milazzo, M.; Selvik, J. (2018): Using the ALARP principle for safety management in the energy production sector of chemical industry. Reliability Engineering and System Safe. 169:160–165.

[2] Arunraj, N; Maiti, J. (2009): A methodology for overall consequence modeling in chemical industry. Journals of Hazardous Materials. 169: 556–574.

[3] Bellamy, L. (2015): Exploring the relationship between major hazard, fatal and non-fatal accidents through outcomes and causes. Safety Science. 71: 93–103.

[4] Concepción-Maure, L. (2018): Support methodology for decisional assistance in the Technological Risk management process. Central University "Marta Abreu" of Las Villas."unpublished"

[5] Fyffe, L. (2016): A preliminary analysis of Key Issues in chemical industry accident reports. Safety Science. 82: 368–373.

[6] Gruden, D. (2008): Umweltschutz in der Automobilindustrie: Motor, Kraftstoffe, Recycling.

[7] Hirst, I.; Carter, D. (2002): A "worst case" methodology for obtaining a rough but rapid indication of the societal risk from a major accident hazard installation. Journals of Hazardous Materials. 92: 223–237.

[8] Johansen, I.; Rausand, M. (2014): Foundations and choice of risk metrics. Safety Science. 62: 386–399.

[9] Li, C.; Ren, J.; Wang, H. (2016): A system dynamics simulation model of chemical supply chain transportation risk management systems. Computers and Chemical Engineering. 89: 71–83.

[10] Marulanda, M.; Cardona, O; Barbat, A.; (2009): Robustness of the holistic seis-mic risk evaluation in urban centers using the USRi. Journal International Society for the Prevention of Hazards. 49: 501–516.

[11] Meel, A.; Seider, W. (2008): Real-time risk analysis of safety systems. Computers and Chemical Engineering. 32: 827–840.

[12] Sujan, M. (2017): How can health care organisations make and justify decisions about risk reduction? Lessons from a cross-industry review and a health care stakeholder consensus development process. Reliability Engineering and System Safety. 161: 1–11.

[13] Tixier, J.; Dusserre, G.; Salvi, O.; Gaston, D. (2002): Review of 62 risk analysis methodologies of industrial plants. Journal. Loss Prevention Process Indusries. 15: 291–303.

MODERN WAREHOUSE AUTOMATION WITH GPS AND RFID

Olga Morozova
Department of Theoretical Mechanics, Mechanical Engineering and Robotic Systems/Engine Design Faculty
National Aerospace University "Kharkiv Aviation Institute", Ukraine

Larysa Pristupa
Department of Theoretical Mechanics, Mechanical Engineering and Robotic Systems/Engine Design Faculty
National Aerospace University "Kharkiv Aviation Institute", Ukraine

Iryna Malykhina
Department of Theoretical Mechanics, Mechanical Engineering and Robotic Systems/Engine Design Faculty
National Aerospace University "Kharkiv Aviation Institute", Ukraine

Effective functioning of any enterprise depends primarily on the well-coordinated functioning of all its departments, from the procurement department to the finished goods department. In addition, it should be noted that of great importance for the successful operation of the enterprise are well-organized activities of its warehouse and transportation facilities. And if the enterprise does not pay due attention to these facilities, they start running rough and that in turn can negatively affect the overall output.

This paper focuses warehouses as enterprise departments that provide efficient goods storage and enterprise stocks transfer. A well-organized warehouse will ensure lower logistics costs.

Consequently, the relevant problem is to find ways of improving warehousing operations [1].

Cargo handling is a set of operations performed at different stages of warehousing [2].

The operations of cargo handling within the warehousing process include four main stages:
- Receipts of Goods (a set of logistic operations performed when goods are received; the operations may vary greatly depending on the type of goods arriving);
- Acceptance of Goods in terms of quantity and quality;
- Stacking and Storage of Goods (the goods are piled up or stacked on pallets);
- Dispatch of Goods from the Warehouse (the operations of goods extraction from their storage places, batching, freight forwarding, vehicle loading).

Table 1 shows the main logistics technologies for cargo handling optimization.

Thus, the warehousing process includes the above operations that, if fulfilled efficiently, are a key factor in warehouse performance. Hence, of great importance is to improve both individual materials handling operations and the functioning of the entire warehouse.

Individual cargo handling operations	Warehouse operation optimization
Barcoding, radio-frequency identification	Warehouse Management System (WMS)
Cross-docking	Simulation Modeling
ABC-XYZ-analysis	GPS

Table 1 – Logistics technologies for cargo handling optimization

One of the modern trends in warehouse automation is the technology of warehouse management system (WMS). This technology allows one to significantly reduce the runtime of the operations, their cost and the number of errors. It also improves the quality of customer service, staff productivity, reduces the costs of goods storage. All in all, the WMS provides most efficient warehouse management. The principle of WMS is described below. The warehouse territory is divided into areas by types of technological operations. This allows one to automate the warehouse operations, namely the acceptance, arrangement, and storage of goods, order-picking,

and dispatch of goods. In this way the work of the staff is better organized in different areas, with effective allocation of their responsibilities. At the implementation stage, the system is filled in with the description of physical characteristics of the warehouse, the number of cargo vehicles, the parameters of all the equipment used and the operating instructions. To perform cargo loading and unloading operations the staff are equipped with radio terminals for data input and output. The terminals are portable computers that communicate with the main server of the system through the radio channel.

The system takes into account all the requirements for storing conditions when the storage space is allocated for the goods admitted to the warehouse. For example, humidity and temperature mode, as well as shelf life, manufacturers, sell-by dates, suppliers, and compatibility of goods and any other parameters can be taken into account. The WMS system automatically selects storage places for the accepted goods and specifies the tasks for the warehouse staff.

The paper suggests automating the work of modern warehouses through the use of GPS and RFID systems.

GPS Monitoring is a comprehensive solution that allows one to track and control the vehicles traffic and what happens when they move from one location to another in real time. Besides, the information can be accessed from anywhere through a web interface [3]. The program allows one to improve the efficiency of management of vehicles and lorry fleet, minimize the cost of lorry fleet operation, control the location of the vehicles, and minimize the human factor. Tracking of the vehicles is based on GPS-technologies using trackers and various sensors. The tracks of vehicles movement are displayed on an electronic map.

The GPS monitoring module operates as follows:

• receives and reviews the information on actual transportations;

• informs about the events;
• compares the actual data with the planned data;
• tracks the transportation progress;
• adjusts the transportation plan;
• ensures the feedback from the driver through the message exchange.

Another promising technology in warehousing logistics is the radio frequency identification (RFID). The main elements of the system include tags, an antenna and a computer. The information is placed on the tag. The tag is placed inside the pallet. After that, all the tag information is transferred to the computer through the antenna. The use of radio frequency identification allows one to control the movement of goods, reduce the time of information processing and, thus, reduce the overall costs [4].

The RFID technology has a great advantage since it increases the visibility of objects within the supply chain. It allows the enterprise to manage the goods supply chain successfully and at a higher performance. This technology ensures tracking of goods in real time and, as a result, this leads to a better control of the stocks and quick reaction to the changes (flexible management).

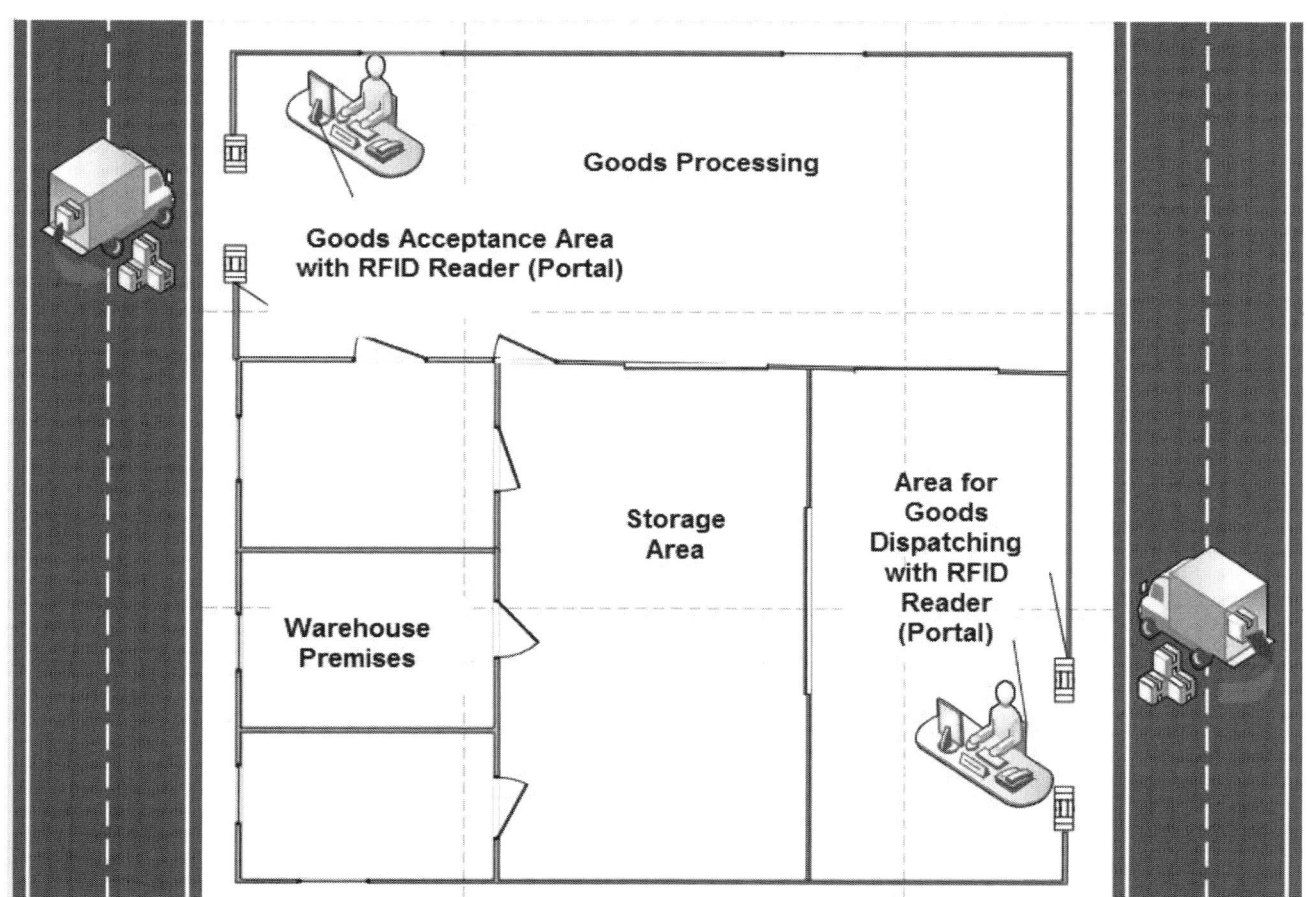

Figure 1 – Warehouse Layout with the GPS and RFID technologies

The RFID systems are divided into the following groups by their read range:
- short-range identification (within 0,2 m);
- mid-range identification (within 0,2-5 m);
- long-range identification (within 5-300 m).

Most RFID tags consist of two parts. The first part is an integrated circuit (IC) for storing and processing of information, modulating/ demodulating of a radio frequency (RF) signal and some other functions. The second part is an antenna for reception and transmission of the signal. The RFID tag is a "radio frequency tag" that allows one to record:
- goods code;
- expiry date, lot number, serial number;
- goods manufacturer / owner.

The RFID technology is well implemented in the inventory management program, more specifically the WMS system. The RFID technology makes warehouse logistics more transparent and warehouse accounting easier.

The development of the RFID system implies equipment selection for the warehouse operations. When choosing the architecture of the system it is necessary to take into account the technical parameters of the equipment and their compatibility.

An RFID portal reader is designed to register a large number of RFID tags within the passages under its control. The reader consists of two vertical bars, each with 2 transceiving antennas. Antennas provide an area for registering RFID tags, completely covering the entire passage between the bars. The walls of the bars are made of radio-transparent material. The RFID portal operates at 220V. It has a system of uninterruptible power supply to neutralize the short-term loss of power supply. The RFID reader can operate autonomously or under control of external software.

The autonomous operation mode of the reader can be used in the library system, when multiple tag data are used to make an independent decision on whether the tag dislocation is autorized or else there will be a sound of warning signal. When controlled by external software, the RFID reader operates using the Ethernet network, the LLRP protocol. The reader bars are attached to the floor with the help of archors [8].

By introducing transport control, one can significantly facilitate logistics processes. As a result, one can obtain the information about the use of vehicles, check their location and fuel consumption whenever necessary. In addition, the GPS monitoring will help automate the compilation of transportation routes.

This system can be effectively employed in vehicle fleets of any size and type. To put it into action, one has to install monitoring equipment and software.

The GPS control offers a range of benefits as follows:
- tracking of vehicle routes, speed modes and goods delivery;
- prevention of vehicle downtime, parking violations (including prohibited geozones;
- better efficiency of logistics services;
- minimal risks of vehicle and cargo thefts;
- prevention of unwanted traffic, deviations from the traffic routes and schedules, deliberate route extension;
- integration with tachographs to provide total control of the vehicles. The GPS system consists of a reliable highly-sensitive tracker, whose main function is to receive and transmit information about a vehicle, a GSM/GPS antenna that receives incoming information and transmits it to the tracker, a satellite navigation module, and a GSM alarm system [5].

One more specific feature of this research is a simulation of goods transportation from the manufacturer to the warehouse by means of the AnyLogic (Personal Learning Edition for beginners and students) package (Fig. 2).

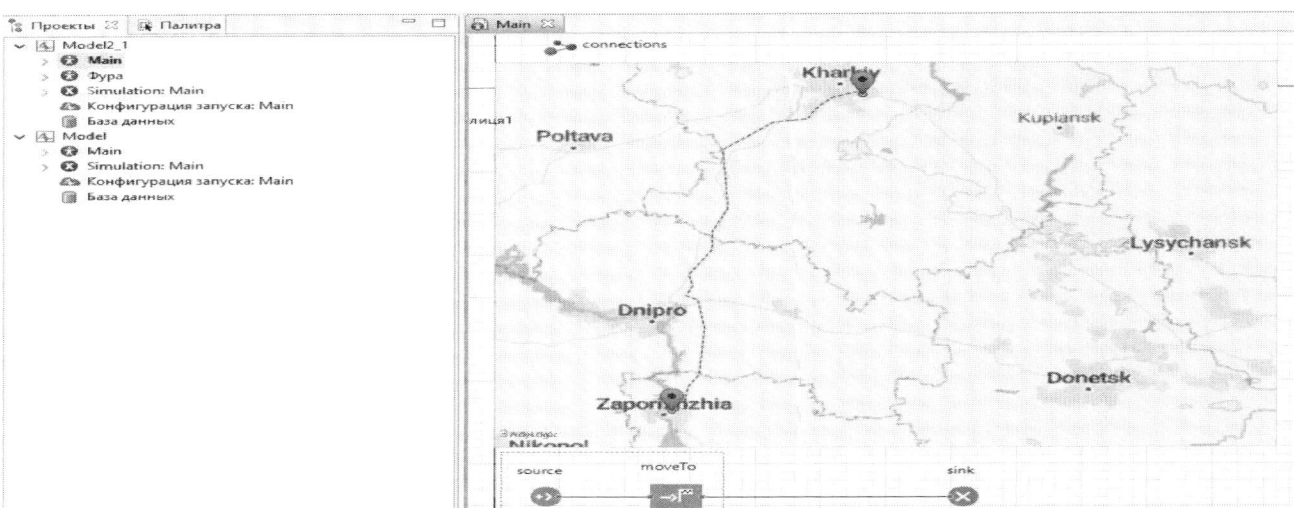

Figure 2 – Model Simulation with AnyLogic (Personal Learning Edition for Beginners and Students

This allows one to plan and choose the optimal route for goods transportation, and assess the risks associated with vehicle downtime (when loading / unloading the goods or crossing the border).

Thus, it is obvious why the GPS and RFID systems are vital for modern warehouse automation. First, the goods entering the warehouse are checked with RFID reader at the portal, where the reader recognizes the location and identification of the RFID tags attached. This process is managed by the operator responsible for the goods acception (Fig. 1). Second, the goods are processed i.e. checked for flawlessness and registered in the database. Then, the goods with tags are transported to the warehouse for storage. Afterwords, the goods will be dispatched to the customer in lots controlled with RFID readers.

In addition, the GPS systems are suggested to be used for cargo tracking.

An operator responsible for goods acceptance has a personal computer at his/her work station that uses special GPS software to track the truckload at every point of the route from one warehouse to the other.

To conclude, the authors of the present paper suggest that modern warehouse automation can be successfully performed through the use of GPS and RFID systems and explore possibilities for introducing these technologies in modern warehouse complexes.

References

[1] Titorenko, G.A. (2011). Informatsionnyie sistemyi i tehnologii upravleniya [Informational Systems and Management Technologies]. 3d Ed. Moscow: YuNITI-DANA.

[2] Dyibskaya, V.V., Zaytsev E.I. et al. (2009). Logistika [Logistics]. Moscow: Eksmo.

[3] Sergeev, V.I. (2008). Logistika: Informatsionnyie sistemyi i tehnologi [Logistics: Informational Systems and Technologies]. Moscow: Alfa-Press.

[4] Rodkina, T.A. (2001). Informatsionnaya logistika [Informational Logistics]. Moscow: «Ekzamen».

[5] Strelnikova, I. A., Artemova, Yu. A. (2013). SPUTNIKOVYiE SISTEMYi NAVIGATsII I MONITORINGA TRANSPORTA [Satellite Navigation and Traffic Control Systems]. *Avtomobil i Elektronika. Sovremennyie Tehnologii,* № 1, *56-62.*

[6] Karpov, Yu. (2005). Imitatsionnoe odelirovanie sistem. Vvedenie v modelirovanie s AnyLogic 5 [System Simulation. AnyLogic for beginners]. Saint Petersburg: BHV-Peterburg.

[7] Kiseleva, M.V. (2009). Imitatsionnoe modelirovanie sistem v srede AnyLogic : uchebno-metodicheskoe posobie [Simulation of Systems with Anylogic]. Ekaterinburg: UGTU – UPI.

ORDER SCHEDULING OPTIMIZATION METHOD OF CONSIGNMENT SELLER DIETARY SUPPLEMENTS MANUFACTURING COMPANIES

Szabolcs Szentesi PhD student
Institute of Logistics
University of Miskolc, Hungary

Péter Tamás PhD
Institute of Logistics
University of Miskolc, Hungary

Béla Illés Prof. Dr.
Institute of Logistics
University of Miskolc, Hungary

Abstract:
Today, the development of competent supply chains plays an important role in preserving the competitiveness of enterprises. The professional international literature examines supply chain design and operations extensively, but there are some types that supply chains which analysis of the development potential has not yet been in focus. One of the major problems with consignment seller dietary supplements manufacturing companies' supply chains is the optimization of order scheduling, as many decisions need to be considered together. Due to the different considerations, a simulation model is needed. The paper presents this problem and the solution method we have developed.

1   Introduction

Today, the expansion of the number of product categories claimed by buyers has implications for the procurement logistics business of the merging companies. It can be said that this also increases the number of types of raw materials to be purchased and the volume of products to be stored, which can be explained by the fact that due to the expansion of the product structure, – to meet customer needs – a minimum set of products has to be stored [1].

Raw materials are purchased by most of the dietary supplement companies from China or USA [2], which can lead to significant lead times of up to several months. Suppliers in the region can source raw materials at a higher cost but shorter lead time. Experience shows that the quality of the raw material also differs considerably, as the quality of raw materials from neighbouring countries can be obtained, such as, for example,

from China [3]. It can be stated that for companies that manufacture food supplements, the ordering of raw materials for a particular product should take place by taking into account a number of factors [4].It means thata simulation test model is neededfor determining when and what to order. The literature does not deal with this area or only minimally, as it focuses only on the acquisitions of large companies [5], so we have set the target for discussion as the optimization of order scheduling can reduce operating costs and improve market competitiveness.

2   Consignment seller dietary supplements manufacturing companies' procurement logistics system

The typical operation of a consignment seller dietary supplements manufacturing companies' procurement logistics system is shown on Figure 1.
Presentation of system objects:
- *Suppliers:* Companies supplying raw materials.
- *Central Base Station:* A warehouse for storing raw materials arriving from suppliers.
- *Manufacturing Plant:* Capsule production is carried out at this location.
- *Finished product warehouse:* The finished products are stored in this warehouse at the capsule production plant.
- *Commissioning area:* In this section, they will pick up and then pack the finished product to be sent as a replacement.
- *Buyer:* The buyer receives the products in a stockist way and produces a weight loss report to the central company about the quantity sold at certain intervals.

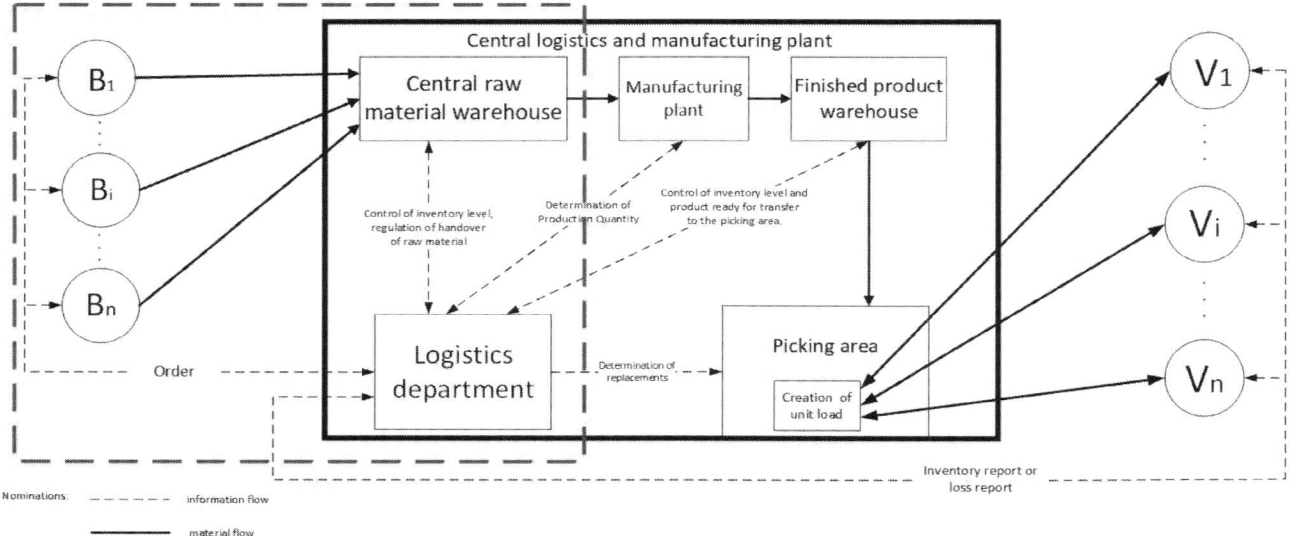

Figure 1: Consignment seller dietary supplements manufacturing companies'material and information flow

Understanding the operation of the system:

In Figure 1, the continuous arrow is the material, the dashed arrow shows the information flow. Suppliers are supplying the raw material required to make the finished product on a predetermined basis, which is replaced by the production system based on customer stock data. In most cases, production is executed on the basis of a Push principle; they are stockpiled, where the product can be several months before it is dispatched to the point of sale. Raw materials that are pre-ordered from suppliers are delivered to the central raw material warehouse where they are stored until they are transferred to the capsule manufacturing department. This is where FIFO is done by storing and delivering the raw material to the manufacturing plant. Preparations of the homogenized base formulation are prepared in the manufacturing plant where they are boxed and labeled after the preparation.

Once a full dose of a homogenized amount is loaded, boxed and labeled, it is delivered to the finished product warehouse. In the finished warehouse is also stored under FIFO principle. The picking department should be kept at a certain product level. From there, the replacement of the products to the buyers of customers happens. Once the number of pieces is replaced, the picking, boxing and dispatching of the products is done individually. In practice, the operation of warehouse management systems (WMS) available for enterprise management (ERP) systems does not always meet corporate needs. In many cases, several development options remain unexploited (eg. optimized material handling, system evaluation functions, etc.) [6]

3   Order scheduling methods for consignment selling companies

When optimizing the orders of companies that produce food supplements, we have to decide which stock management mechanism to choose from, as the inventory mechanism has an impact on the procurement logistics process. Most companies attempt to make their orders lean, but this approach not only provides helpful advice to decision-makers, but also unfortunately challenges them [7]. When ordering, we have to decide on two factors: when and how much to order. The description of inventory that records the quantity and time parameters of the order is called the inventory mechanism.

You can order a fixed amount (q) or a quantity up to the maximum level (S). The stock to be stocked can be ordered at fixed intervals (t) or only at the time the set reaches a specified level (s).

Based on the above combinations we distinguish four basic mechanisms:

- (t, q): order quantity recorded per inventory cycle;
- (t, S): uploads to a maximum inventory level at fixed intervals;
- (s,q): Order a predetermined quantity (q) at the specified set level;
- (s, S): fills up to the maximum inventory level when reaching the set inventory level.

Basically, two useful and relatively easy-to-manage inventory mechanisms and their combinations were used in company practice [8]:
– A continuous review system where inventory levels can be practically measured at any moment of operation and based on the characteristics of the current moment we can make a decision when needed. In this case, the decision variables define the order quantity (q) and the inventory set (s),

i.e., the inventory level at which the order is assigned to the predetermined amount. This inventory mechanism is called (s, q) mechanism.

– In the case of a periodic review system, the inventory level is examined only at certain fixed intervals (e.g. weekly, monthly), and on this basis, we decide on the amount of order quantity. The order quantity shipped with this mechanism varies in time and depends on the size of the actual inventory measured in the review period. The order quantity in this case is the difference between the possible maximum inventory level (S) and the actual inventory size measured at the time of observation. This mechanism (t, S) is a stock mechanism [9].

However, in the case of the type of company under investigation, most of the time the (s, q) mechanism is used, as the vendors and their sales in the commission network change stochastically. For this reason, we need to be able to look at inventory levels at any moment and, if necessary, make a decision based on the characteristics of that moment.
For the inventory (s, q)mechanism, the reordering point (or reorder level) policy is chosen.

Per experience, the inventory level (s, q) of inventories stored by the companies that manufactures food supplements is usually set uniformly for the order stock level (s) and the order quantity (q) for each product type. By developing and implementing a simulation test system, it would be possible to determine the s and q values for each type of product:
- the fulfilment of production needs,
- the lead time of the procurement logistics process,
- purchasing logistics costs,
- stochastic effects (delay in delivery, poor quality product delivery, etc.). This would increase the reliability of procurement logistic processes and / or reduce the running costs.

In the following, the operating concept of this test system is described.

4   Presentation of Optimization Method

The (s, q) inventory mechanism described in the previous chapter is mostly based on calculations based on intuitions and simple calculations that consider only a few aspects, which in most companies causes significant losses. To minimize these losses, it is necessary to develop a simulation application, the starting concept of which is described in this chapter. In the simulation model, all (s, q) inventory mechanism variations should be generated for all product types, which must be executed by selecting the variation where the value of the target function is most favourable. The target function is the purchase logistics cost.

During the simulation test, all possible (s, q) inventory mechanism pairs must be established for each product type, and then their variations must be trained (one variation for a product type has a stock exchange mechanism value pair). If we run the simulation for each specified variation, we can select the best value function (product types and the associated stock selection mechanism pair selection) so that always the selection of the inventory mechanism parameters that best suits the company's interests will be selected and applied, thus significantly reducing logistics losses. The value of the target function is the cost of the entire procurement logistics system.

Due to the simulation test time, if the number of products is low, then all variations are examined, but if the number of products and related ingredients is high, then a genetic algorithm may be needed to handle run time.

The steps for the simulation test are shown in Figure 2.

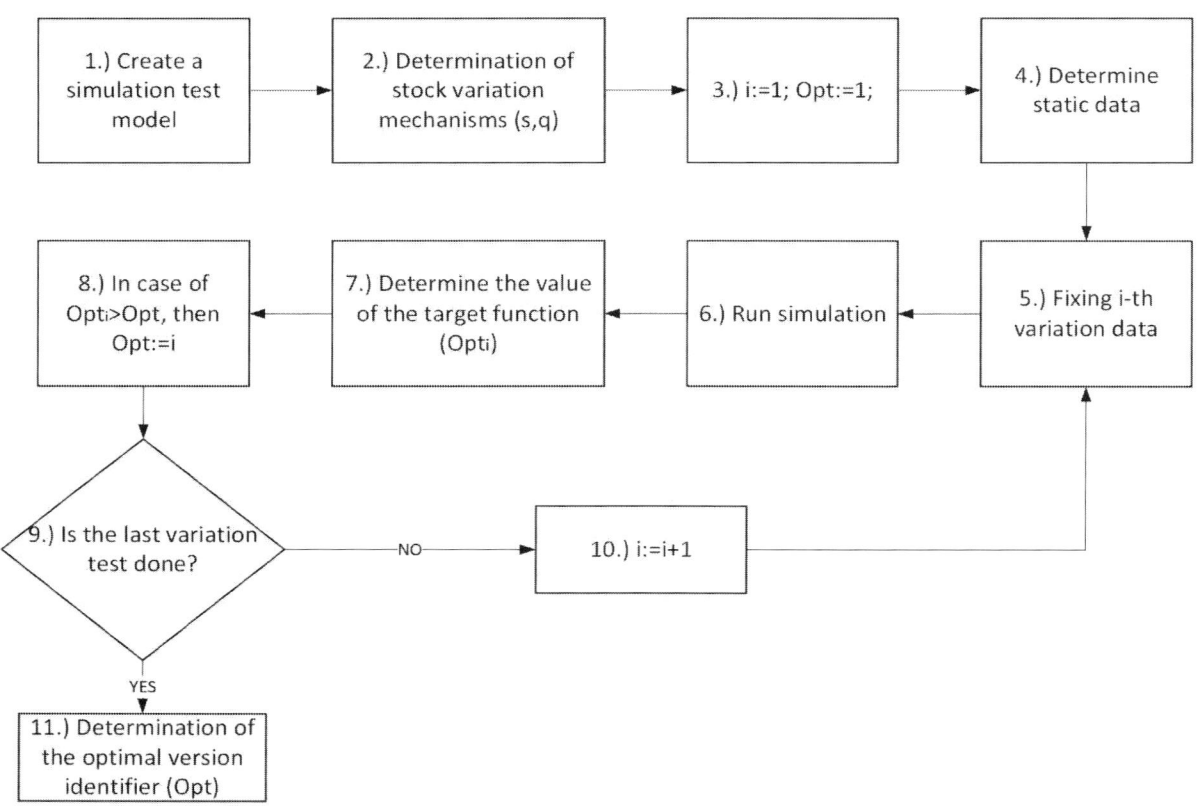

Figure 2: Optimum (s, q)inventory mechanism concept of choice of variation pair variation

Steps for selecting the value pair variation of the optimal inventory mechanism:

Step 1: Create a simulation test model: The preparation of the simulation test and the implementation phases are as follows:
- Definition of the purpose of the simulation, limitation of the examined logistic system.
- Understanding the operation of the tested system.
- Determine the set of logistical indicators needed to achieve the test goals.
- Define input and output data.
- Create a simulation model.
- Checking and repairing a developed model.
- Evaluation of test results, formulation of suggestions.

Step 2: Determination of stock variation mechanisms (s, q).
We determine the possible stocking mechanism value pairs by product type and then form their variations.

Step 3: Initialize Variable Values:
The variable i is the (s, q) inventory mechanism pair variation that is being investigated, and Opt is the optimum inventory quantity identifier.

Step 4: Determine and record static data.
Record the data needed to run the simulation.

Step 5: Determine the value of the i function of the i-th version, the target function is the purchase logistics cost:

Step 6: Run Simulation

Step 7: Determine the value of the function of the i-th version using an optimized inventory mechanism.

8. Investigate whether the target function value of the i-th stock option is more favourable than previously defined. If so, the variable Opt will be stored in this variant identifier.

Step 9: Investigate whether the last inventory variant has been tested.

Step 10: If the last inventory check has not been done, the value of the variable i must be incremented so that the next inventory variant should also be executed in the 6-8. steps.

The paper presented is an initial concept of an initial research, and I will later concretize specific steps.

5   Summary

One of the most commonly used forms of access to the market for food supplement manufacturing companies is commission sales, which is a mutual benefit for both manufacturing and sales companies. The ever-expanding market has brought about the diversification of customer needs, which have become more and more difficult to inventory. In the thesis, we presented the operation of the typical purchasing logistic system of the food supplement companies, as well as a possible method of optimizing the value pairs of the stocking mechanism (s, q) by product type. The use of the method can result in significant benefits to dietary supplement companies, as it may significantly reduce purchasing logistics costs.

6 Acknowledgements

This project has received funding from the European Union's Horizon 2020 research and innovation programme under grant agreement No 691942. This research was partially carried out in the framework of the Center of Excellence of Mechatronics and Logistics at the University of Miskolc.

References

[1] Peres, R., Müller, E., Mahajan, V. (2010): Innovation diffusion and new product growth models: A critical review and research directions, International Journal of Research in Marketing, Volume 27, Issue 2, June 2010, pp. 91-106.

[2] Behrens, A., Giljum, S., Kovanda J., Niza, S. (2007): The material basis of the global economy: worldwide patterns of natural resource extraction and their implications for sustainable resource use policies. Ecol. Econ. 64 (2), Special Section—Ecosystem Services and Agriculture Ecosystem Services and Agriculture, pp. 444–453.

[3] Szentesi. Sz., Illés, B., Tamás, P., (2017): Supply Chain Improvement Possibilities of Consignement Seller Dietary Supplements Manufacturing Companies In, Michael Schenk 10th International Doctoral Student Workshop on Logistics, pp. 143.

[4] Vörösmarty, Gy., (2002): A beszerzés információs kapcsolatai, PhD értekezés, Budapesti Corvinus Egyetem

[5] Arjan, V., (2005): Purchasing and Supply Chain Management, 4th ed. Thomson Learning, London

[6] Tamás, P., Illés, B., (2016): Raktár-irányítási rendszerek hatékonyságnövelési lehetőségeinek vizsgálata, MŰSZAKI SZEMLE (EMT) 68, pp. 29-37.

[7] Molnár, V., Kerchner, A. (2016): A lean menedzsment alkalmazási lehetőségei a közszférában. Műszaki tudomány az Észak-Kelet Magyarországi régióban, pp.799.

[8] Chikán, A. (1976): Tartalékok a vállalati rendszerben. Ipargazdasági Szemle különszám, pp. 108-114.

[9] Dobos, I., Gelei, A. (2015) Biztonsági készletek megállapítása előrejelzés alapján - Esettanulmány.Vezetéstudomány. Budapest Management Review, 46 (4). pp. 14-22.

REDUCTION OF THE BULLWHIP EFFECT IN THE CHAIN OF SUPPLY OF THE COMPANY LABIOFAM VILLA CLARA WITH A MODEL VENDOR MANAGEMENT INVENTORY (VMI)

Ing. Ernesto González Cabrera
Industrial Engineering Department, Central University from Las Villas, Cuba

DrC. Roberto Cespón Castro
Industrial Engineering Department, Central University from Las Villas, Cuba

Prof. Dr.-Ing. Dr. h.c. Norge Isaias Coello Machado
Mechanical Engineering Department, Central University from Las Villas, Cuba

Dr.-Ing. Dr. h.c. (UCLV) Elke Glistau
Institute of Logistics and Material Handling Systems
Otto von Guericke University Magdeburg, Germany

Abstract

With the purpose of diminishing the impact of the bullwhip effect in the link of the supply chain between the warehouse and the points of sales is carried out the present investigation in the Company LABIOFAM Villa Clara. The investigation is elaborated based on a bibliographical analysis in which aspects related with the contents study object are approached to fulfill the proposed objective. In the development of the investigation, leaves of an analysis to determine the product that more impact has in the storage costs, as well as its rotation. Then their current situation is determined and it is compared with the selected pattern that it follows a philosophy Vendor Management Inventory (VMI). Also, the impact of the bullwhip effect is determined in the current pattern and in the one proposed to confirm the impact and to facilitate the taking of decisions; although to supplement this analysis they are also kept in mind the possible new costs that associate brings the new model that intends. To arrive to the obtained results it was necessary to use technical of demand presage, determination of the size of the lot, interviews with personal of the center, direct observation and it consults to documents of the entity those that offer a scientific support to the investigation. After having applied the procedure it was demonstrated that the company can reduce the impact of the bullwhip effect, starting from the implementation of the proposed pattern that it follows the philosophy VMI.

Keywords: bullwhip effect, supply chain, Vendor Management Inventory.

1. Introduction

The current business world is becoming increasingly complex and unpredictable for companies globally. The increase in the competition of certain productions, the economic, financial, energy, food and environmental crisis and the accelerated development of science and technology, together with the globalization of the market, mean that all organizations, especially Cuban ones, face a race to find solutions that assure them a position in the market, help them to optimize their processes and make them more competitive.

Cuban companies do not escape the impact of this global scenario, which is why they have adopted new competitiveness strategies to consolidate their project. These strategies are geared to meet the expectations of customers in a growing manner, so that they are offered better products and services every day that offer greater opportunities and lower costs. Therefore, the mission of the top management is directed to direct its actions under the concept of incessant changes, at accelerated rates, with great complexity, imposing in this way the need for flexible solutions to the difficulties, on a scientific basis.(López Quintero, 2014)

For the national industry it is essential to apply new ways of managing the company, such as the focus on systems, process management and the integration of flows between suppliers and customers through the direct supply chain. All with the aim of increasing efficiency and effectiveness in organizations, making the most of the installed capacities, reducing production costs and reducing inventory.

The country's policy in recent years has focused on the development of biopharmaceutical and hygiene products. In this sense, the gradual substitution of imports for products of national production of recognized quality has been proposed. The LABIOFAM Villa Clara Company is one of the most important companies in the country that makes this new policy possible.

On the other hand, senior management is interested in the implementation of inventory management systems that allow a better alignment of the company's supply chain with suppliers and customers. The elements contained in this paragraph constitute the problematic situation of this work.

The alignment of supply chains with suppliers and customers and their effect on the reduction of inventories, is known in logistics as bullwhip or efect. In the warehouses of the company are high levels of goods, which involve high storage costs, although this does not occur in all products, raw materials, and inventories in process. This situation is reflected in the low turnover and, consequently, in the productive efficiency of the organization.

Starting from the above, it is defined as a research problem: the use of the appropriate inventory management system will allow the reduction of the whip effect in the supply chain studied and the rotation of the company's inventories.

For the solution of the research problem is defined as a general objective:

Implement an inventory system that allows the reduction of the effect of the whip and the increase of rotations in the warehouses of the selected company as an object of practical study.

To fulfill the objective, the research will be structured in:
1. **Introduction**
2. **Research background.**
3. **Methodology to determinate the inventory level.**
4. **Results**

Containing the selection of the inventories to be studied, the application to them of the inventories system without the VMI and with the VMI; as well as measuring the impact of its adoption on the whip effect of the supply chain analyzed.

2. Research background

It is not a secret that each time the suppliers are required to be more efficient when delivering the goods both in time and in safety; which has led to an increase in models and tools that allow their control and monitoring. Nowadays, with the new technologies, logistics has not been left behind and customers can know where in the world the requested goods are located 24 hours a day and other information thanks to satellites and the internet.

There are many authors and entities that in one way or another have dealt with the concept of logistics, either as Business Logistics or Supply Chain Management, some of the definitions, which demonstrate the evolution of the term through time together with the advance of technologies giving way to new concepts and methods and philosophies.

The extreme competitiveness that exists in the current economy, together with the effects of globalization, force the industry to find new ways to interact and satisfy customers. In a supply chain (CdS), manufacturers, commercial intermediaries, carriers, suppliers and official agencies collaborate to deliver the goods quickly and efficiently so that money flows through the economy.

The bullwhip effect is a phenomenon that hinders administrative management both inside and outside the supply chain and consists of a growing distortion of the demand transmitted by the different agents involved in the management of the flow of products as we move away From the market. In other words, the Bullwhip effect reflects the increase in uncertainty as the orders are transferred upstream in the CdS, in this sense, this effect is considered as the phenomenon of "amplification" of demand, known among the different elements that make up a particular CdS. (Mejía Villamizar, et al., 2014)

Authors such as Holström, Disney, Chen and Kaipia, who have proposed techniques in order to eliminate the whiplash effect and the fulfillment of the real demand through the leveling of inventories such as the smoothing of demand, and the most effective Until now Vendor Managed Inventory.

Vendor Managed Inventory (VMI), It is a system that allows to make the supply chain more agile, by managing inventory levels by the manufacturer or retailer. The typical model of material requirements is given by traditional processes in which the buyer or retailer establishes a demand and the company plans the sourcing processes, reorder points, production planning, inventory levels, etc. On the other hand, with the application of the VMI model, it is possible to reduce delivery times, greater reliability in shipments and reduce transport, production and order costs, which leads to improved production and shipment scheduling, resulting in a greater profitability of the supply chain. The application of a VMI policy will depend on the incentives that each party has to cooperate; for this reason, if one of them has market power, it would not have any motivation to accept a joint policy different from its own and optimal for it (Montenegro Carrascal, 2011).

Action reward learning

The "action reward learning" model, proposed by (Kwak, 2009), is based on determining the replenishment quantities (inventory replenishment), through an analysis of a compensation factor, which determines the minimum cost associated with making a decision of replenishment, based on the amount of adjustment of the replenishment order that minimizes the costs of maintaining inventory or losing income due to not satisfying the total demand, having as a point of comparison the effect of using the compensation factor for the data of the previous period.

3. Methodology to determinate the inventory level and the bullwhip effect.

In this step we calculate the Bullwhip effect in the supply chain. To achieve this goal we will follow the following procedure:

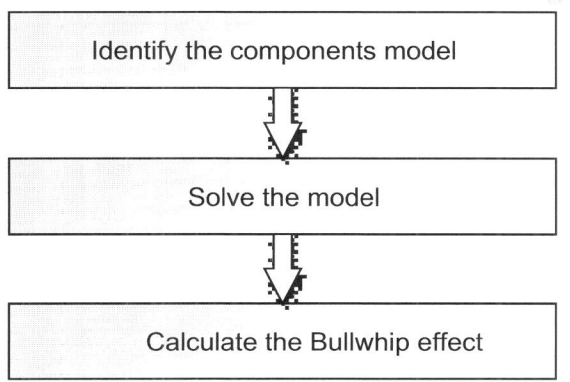

Figure 1: Methodology to determinate the Bullwhip effect

3.1 Identify the componets model

In the realization of this investigation the logistic and economic situation of the company is analyzed, for it is made the analysis of the economic and physical plans of the company, accounting summaries and the cost cards.

The objective of this analysis is to determine the components and find the necessary data to be able to apply the selected model and have economic and financial data to compare.

3.2 Solve the model Action reward learning

The algorithm of the model can be summarized in the way indicated in table 1.1, where the notation of the variables and factors of the model are consigned.

symbol	Description
(t)	Replenishment period (t= 0,1,2,…)
Dt	Actual demands of customers made during [t, t+1)
Dt'	Anticipated step anticipated in the client's demand in period t
ɗt'	Estimated standard deviation for consumer demand in period t
It	Inventory level at the beginning of period t
Qt	Amount of replacement at the beginning of period t
ρi	Value of the CF compensation factor
Θ	CF value set (Θ= (ρ1, ρ2,…, ρ n))
h	Cost of maintaining inventory by SKU (Stock-Keeping Unit)
S	Cost due to shortage of
	inventory by SKU
Ct(ρi)	Costs incurred in inventory in period t, when CF (ρi) is chosen in period t-1
Ct(ρi)'	Average inventory cost calculated in period t, when CF (ρi) is chosen in period t-1.

Table 1.1. Notation for the model action-reward learning Source :(Arango, 2011)

The inventory level of retailer it at the beginning of the replacement period t is calculated by equation 1.

$$It = It\text{-}1 + Qt\text{-}1 - Dt\text{-}1 \quad t = 1, 2, 3,\ldots \quad (1)$$

The cost of maintaining the inventory of each SKU is h and it occurs when It is positive. The cost per shortage of each SKU is s and occurs when It is negative. The replacement quantity Qt, at the beginning of period t, consists of an average projection of the demand and an addition (or subtraction) produced by the compensation factor (CF), then:

$$Qt = Dt' + (1+\rho i)\, ɗt' - It \quad (2)$$

Where ρi is the chosen value of CF in the replacement period t; It can be negative or positive. ɗt 'is the estimated standard deviation of customer demand, calculated as ɗt' ≈1.25 x MADt. MADt represents the average absolute deviation of the error forecasts; it is calculated as:

$$MADt = (1\text{-}\gamma)\, MADt\text{-}1 + \gamma|Dt\text{-}1 - Dt\text{-}1'|.$$

The inventory cost that occurs at the beginning of period t by ρi is calculated as:

$$Ct(\rho i) = \varepsilon \times |It|, \text{ donde } \varepsilon = (x = h \text{ si } It >= 0 \text{ ó } x = s \text{ si } It <= 0) \quad (3)$$

The cost Ct (ρi) 'is averaged with the previous values and is denoted as Ct (ρi). The average cost of inventory for a value of CF ρi is found using the exponential method of the weighted average defined in the equation 4.

$$Ct(\rho i)' = Ct\text{-}1(\rho i)' + \beta ti\, [Ct(\rho i) + Ct\text{-}1(\rho i)'] \quad (4)$$

Where C0 (ρi) = 0 for pi ε Θ

βti is an adaptive smoothing parameter. When client demand changes abruptly, βti must take a high value, so that the recent demand data have more weight in the equation of the total average cost (equation 5). When the demand behaves in a stable manner, βti must take small values. The parameter is calculated with the equation 5.

$$\beta ti = |MDti/MADti| \quad (5)$$

According to Kwak et al. (2009), the range of the value of CF can be considered as a necessary parameter for "action-reward learning". The same authors suggest using CF values between 2 and -4, which covers almost any demand, since it ensures that the model takes the average demand of the consumer, more than three times the standard deviation above and below said half.

The best CF value (ρ *) must be selected with the value of ρ * that minimizes costs. The way to perform the selection of said parameter is explained in the original work of Kwak. (2009). (Arango, 2011)

3.1 Calculate the Bullwhip effect

The mathematical measure of the whip effect in terms of the oscillation of the demand that a CdS experiences, relates to the squares of the coefficients of variation of the transmitted and received demand, starting from the assumption that in the medium and long term the average values of the demands. (See equation 6) in figure 2.

$$BW = \frac{\dfrac{Var(q)}{d_q^2}}{\dfrac{Var(d)}{d_d^2}} = \frac{Var(q)}{Var(d)} \quad (6)$$

Figure 2: Whip effect equation. Source: (Mejía Villamizar, et al., 2014)

Where:

➢ Var (q) and Var (d): Correspond to the variances of both demands, the one transmitted by the demand that the customers of the company expect to satisfy and the one received from the actual demand that they satisfy, respectively, in said step of the chain of supply.

➢ dq and dd: Are the average demands transmitted q, and received d.

➢ BW: It is the indicator used to measure the whip effect produced by a step of the chain, which, when it is greater than 1, means that the transmitted demand is greater than what is actually being sold, evidencing the whip effect. Also if the BW is equal to 1 it means there are no problems; but that is less than 1 does not only mean that the expected demand is lower than the one received, but that the company can fall in costs due to lack.

The formula allows to take the effect of the whip to a mathematical value in order to be able to use it for comparison, analysis and later decision making.

4. Results

In this session will show the results according de details of the methodology in the previous session. This model was applied in the company LABIOFAM Villa Clara, in the production's process of honey with propolis 240ml.

Table 1.2 shows a summary of the main indicators and variables used in the model. The fifth quarter analyzed brings for the first time the missing characteristic of some 9789 units which represent a great cost to the company of $ 75962.64 and are also a cost of opportunities with which you can lose customers.

t	Dt	Dt' pronost	It	Qt	Ct(ρi)
0	6034,00	15000,00	0,00	15000,00	0,00
1	7800,00	15000,00	8966,00	17147,00	2062,18
2	5800,00	15000,00	18313,00	8312,00	4491,21
3	4056,00	15000,00	22569,00	1508,00	5190,87
4	10536,00	15000,00	13541,00	10123,00	2328,41
5	33453,00	20000,00	3211,00	12835,00	738,53

Table 1.2: Summary of the results of the action reward learning model - Source: self made.

This model manages to reduce the inventory costs in the company and the rotation of the product increases to 6.241 times of 3.1677; but it must be borne in mind that the possibility of missing appears, its cost being much greater than that of storing the products.

The costs of maintaining inventory can be observed summarized in table 1.3 with the current model and with the action reward learning model following a philosophy of VMI.

t	Model without VMI		Model wit VMI	
	Inventory level	Costs	Inventory level	Costs with VMI
1	6295	1447,85	8966,00	2062,18
2	8266	1901,18	18313,00	4491,21
3	44930	10333,9	22569,00	5190,87
4	36481	8390,63	13541,00	2328,41
5	9631	2215,13	3211,00	738,53
Total		24288,69		14811,20

Table 1.3: Summary of inventory costs with the different models. Source: self made.

As can be seen by analyzing the costs until December 2017, the model that applies the VMI philosophy is more economical, since it would save only $ 9477, 49 in storage, without counting the cost of producing more units. In addition, when comparing the rotation of the inventory of the two models which are: for the current model of 3.1677 times and that of the proposed model is 6.241 times; It can be seen that this indicator doubles from the current model to the one proposed.

Calculation of the Bullwhip effect

By applying formula # 1 of chapter 1 we can give a numerical value to the whip effect which is used to know how distorted the demand is within the steps in the current and proposed model.

For the current model:

The variance transmitted Var (q) = 5 000 000 will be for the forecast of demand that the company is currently pursuing, which has an average of 16,000 units.

The variance received Var (d) = 145 552 109 is the real demand that the company has satisfied in the periods analyzed, and that has an average of 12 329 units.

BW = (Var (q) / dq2) / (Var (d) / dq2)
BW = 0.01953 / 0.95755
BW = 0.0204

For the proposed model:

The variance transmitted Var (q) = 85 848 562 which is the variance of the amount that should have been requested to apply the model in that time period with an average of 12 585 units.

The variance received Var (d) = 145 552 109 which is the actual demand that the company has satisfied in the periods analyzed, which has an average of 12 329 units.

BW = (Var (q) / dq2) / (Var (d) / dq2)
BW = 0.5420 / 0.95755
BW = 0.566

As shown in the proposed model, the reduction in the impact of the whip effect is significant because it is closer to 1. Although the indicator is still below 1; which means that the actual demand varies much more than expected. The impact is reduced by the model, since the quantities requested are closer to this reality, making its implementation favorable for the company. In addition, the whip effect is affected only with the implantation of the model about 27 times; so that to apply these measures, those proposed in chapter one, this should continue to decrease for the benefit of the company.

5 Conclusions

- The bibliographic search carried out for the preparation of the theoretical-referential framework of the research revealed the existence of a broad updated conceptual base on the subjects under study, as well as previous experiences in the application of a procedure that considers the use of the VMI philosophy to face the bullwhip effect in the supply chain.

- The procedure applied allows to reduce the inventory level in the warehouses and the costs of the selected product; it also increases its turnover, which increases the positive impact within the company.

- The action reward learning model reduces the effect of the whip on the selected step about 27 times what is currently happening in the company; so its application is a feasible solution.

- The results confirm the importance of applying this model in the company to reduce costs so that it can be economically sustainable.

- It is interesting the comparison with other inventory management systems regarding the reduction of the whip effect and analyzing the impact on product quality or other dimensions of competitiveness.

6 Bibliografia

[1] Arango, M. D. (2011). "Aplicación del modelo de inventario manejado por el vendedor en una empresa del sector alimentario colombiano." en *Revista EIA Escuela de ingeniería de Antioquia, Medellín (Colombia)*, Número 15, Febrero 2016, pp. 21-32.

[2] Cespón Castro, R. (2014). *Administración de la cadena de suministro. Manual para estudiantes, académicos y empresarios vinculados al campo de la Logística.* [En línea]. disponible en: https://www.researchgate.net/publication/265963575 [Accesado Febrero 2016].

[3] Holström, J., Smaros, J., Disney, S. & Towill, D. (2003). "Collaborative Supply Chain configurations: the implications for supplier performance in production and inventory control." en *Octavo Simposio de Logística, Sevilla, España.*, Marzo 2016.

[4] Kwak, C. (2009). "Situation reactive approach to Vendor Managed Inventory problem." *Expert Systems with Applications,,* [En Línea]. disponible en: www.elsevier.com/locate/eswa [Accesado Febrero 2016].

[5] López Quintero, L. (2014). *Programación de la producción de artículos elaborados a partir de residuos sólidos en la Agencia Gráfica Offset de la Empresa GEOCUBA VC – SS.* Tesis de Grado.

[6] Mejía Villamizar, J. C., Palacio León, Ó. & Jaimes, W. A. (2014). "Efecto látigo en la planeación de la cadena de abastecimiento, medición y control." en Febrero 2016.

[7] Montenegro Carrascal, M. V. (2011). 'Coordinación de existencias mediante la administración de inventarios por parte del proveedor - VMI.", [En Línea]. disponible en: http://imt.mx/archivos/Publicaciones/PublicacionesTecnicas/pt293.pdf [Accesado Febrero 2016].

SIMULATION AND OPTIMIZATION OF TRAFFIC CAPACITY WITHIN A SEGMENT OF CITY TRANSPORTATION INFRASTRUCTURE

Olga Morozova
Department of Theoretical Mechanics, Mechanical Engineering and Robotic Systems/Engine Design Faculty
National Aerospace University "Kharkiv Aviation Institute", Ukraine

Vadim Vasilyuk
EOS Data Analytics, Ukraine

Tetiana Pavlenko
Department of Economic Theory/Economics and Management Faculty
National Aerospace University "Kharkiv Aviation Institute", Ukraine

Volodymyr Polovynko
Department of Theoretical Mechanics, Mechanical Engineering and Robotic Systems/Engine Design Faculty
National Aerospace University "Kharkiv Aviation Institute", Ukraine

Transportation is the most important part of the country's industrial infrastructure. Its sustained and effective functioning provides a necessary condition for high and stable rates of economic growth of the country, ensures integrity and national security, as well as national defense capabilities, higher living standards of the citizens, and rational integration of the country into the world economy[1-2].

The country's transport system includes road, air, rail, sea transportation, as well as inland waterways and pipelines. The national transportation system consists of the transport infrastructure of each individual city within the country. Besides, the transportation infrastructure of the city comprises industrial transport and urban public transport (Fig. 1).

The lack of a backbone transport network throughout the country can hinder the development of a single economic space and the growth of its citizens' mobility. The growth of the country's citizens' mobility is impossible without a high level of motorization of the country and may be restrained by insufficient development of the highway networks [3-4].

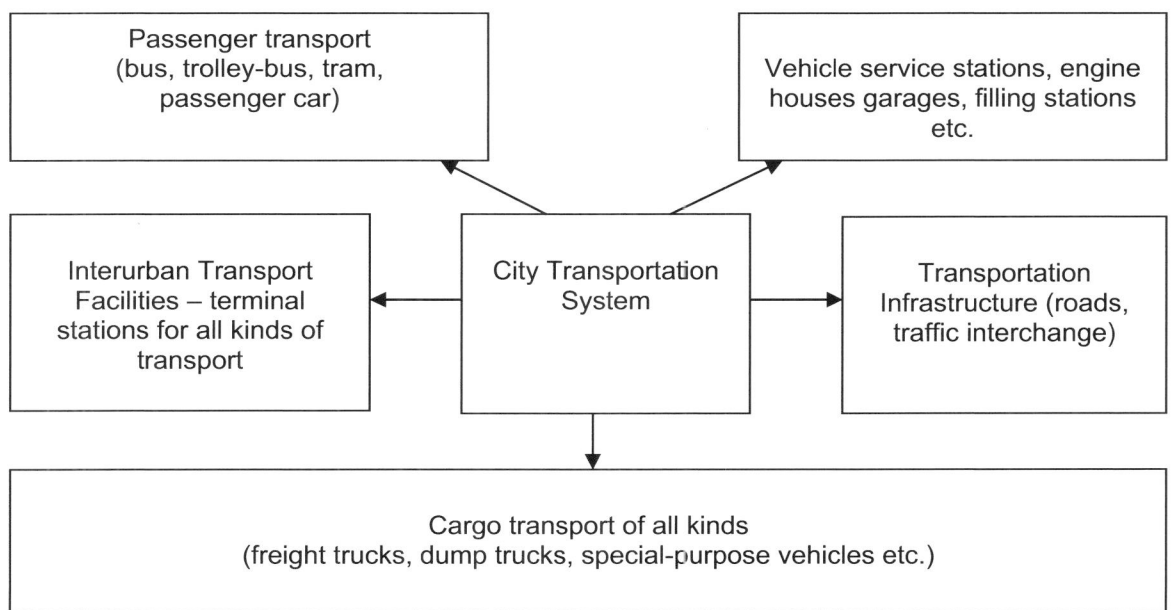

Figure 1 – City Transportation System Layout

Hence, a well-functioning transportation system of the city enables, in its turn, a well-functioning transportation system of the whole country.

When the city's transportation system is developed and starts functioning, the main task is to coordinate its individual subsystems, taking into account the needs of the citizens and all sectors of the economy. More specifically, the parameters of the backbone transport network should correspond to the traffic capacity.

Figures 2 and 3 show the diagrams of traffic congestion during a day and during a week respectively.

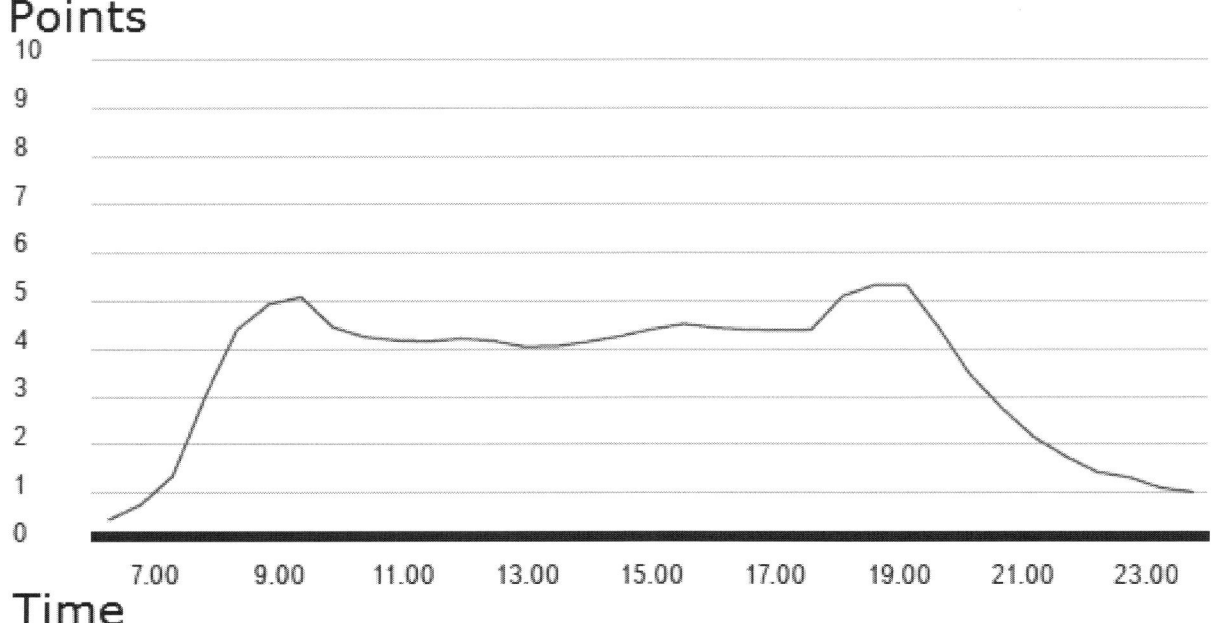

Figure 2 – Traffic Congestion Diagram during a Day

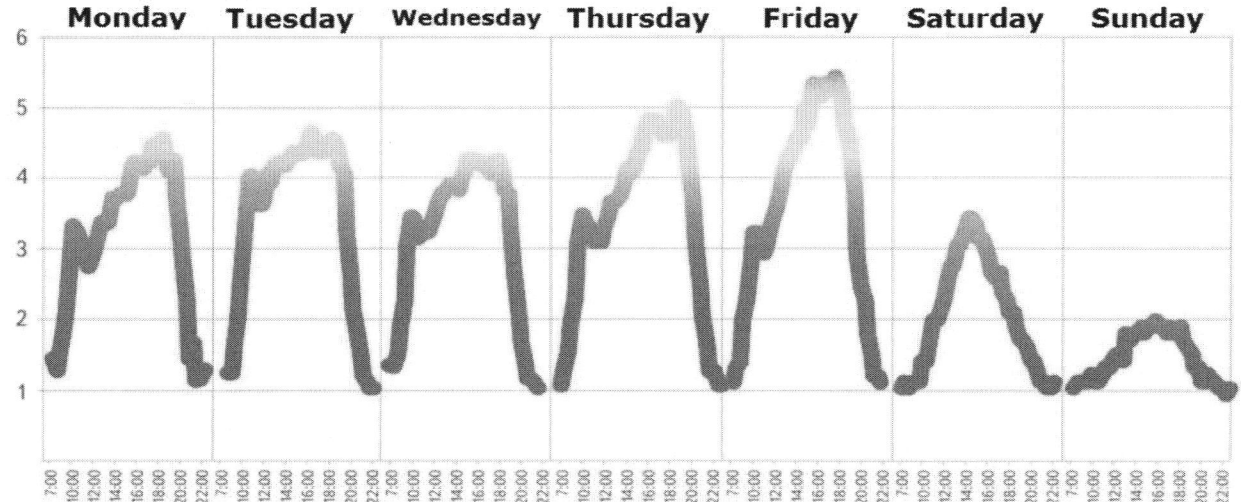

Figure 3 – Traffic Congestion Diagram during a Week

The paper focuses on modeling [5-6] and optimization of the traffic capacity of a segment within the city transportation infrastructure with a case study of Kharkiv City (Ukraine).

The City Transport Planning (CTP) is a complex system with a number of subsystems, namely, a backbone transport network and facilities, railway vehicles, railway depot, garages, parks and repair facilities, as well as a management subsystem. The CTP also includes a personnel resource that operates and maintains the entire transportation system.

As for the transportation industry, the main purpose of the CTP system is to increase its productivity. Given the specifics of transport means production, one should admit that the productivity criterion is the average travel speed of one passenger or one ton of cargo.

Presently, efforts made to improve the city's transportation system can contribute to the main aspects of the city's development, including:

1) preservation of the unity of the largest cities in terms of their development (urban development aspect);

2) reducing the time spent by citizens on travel each day, travel fatigue and a number of accidents, as well as making life in different parts of the city more comfortable (social aspect);

3) decrease of the street and road traffic overloads caused by motor transport and creation of the reserve CTP capacity (transportation aspect);

4) more healthy urban environment as a result of decreased traffic intensity, and consequently, the levels of noise and ambient air pollution (environmental aspect);

5) more intensified use of urban territory due to greater building density, underground urbanization, and energy saving (economic aspect).

One of the important and complex problems of the city's transportation infrastructure is to coordinate transport modes at different territorial levels.

The direct interaction occurs in the transport nodes where different transport modes "contact" with each other. In turn, the interaction of different transport modes consists in the coherence and organization of operations in certain parts of the city.

Today, the aggravation of transportation problems in Kharkiv City is caused by:
• greater number of cars owned by the citizens;
• low density of the main streets and insufficient development of the network of local streets;
• low street capacity;
• combined traffic of public, private and freight transport vehicles;
• poor provision of parking places.

The research deals with the central part of Kharkiv City as a case study. The traffic capacity in this part of the city has been simulated and optimized using the AnyLogic package (Personal Learning Edition for beginners and students) [7-8].

The total length of the road segment selected for simulation experiment is 5.5 km. There are 9 traffic lights wthin the segment (Fig. 4), whose operation mode is shown in Table 1. The traffic lights within the section will be divided into three groups including the blue, orange and green one. For each group, the operating time for the traffic lights is measured in seconds.

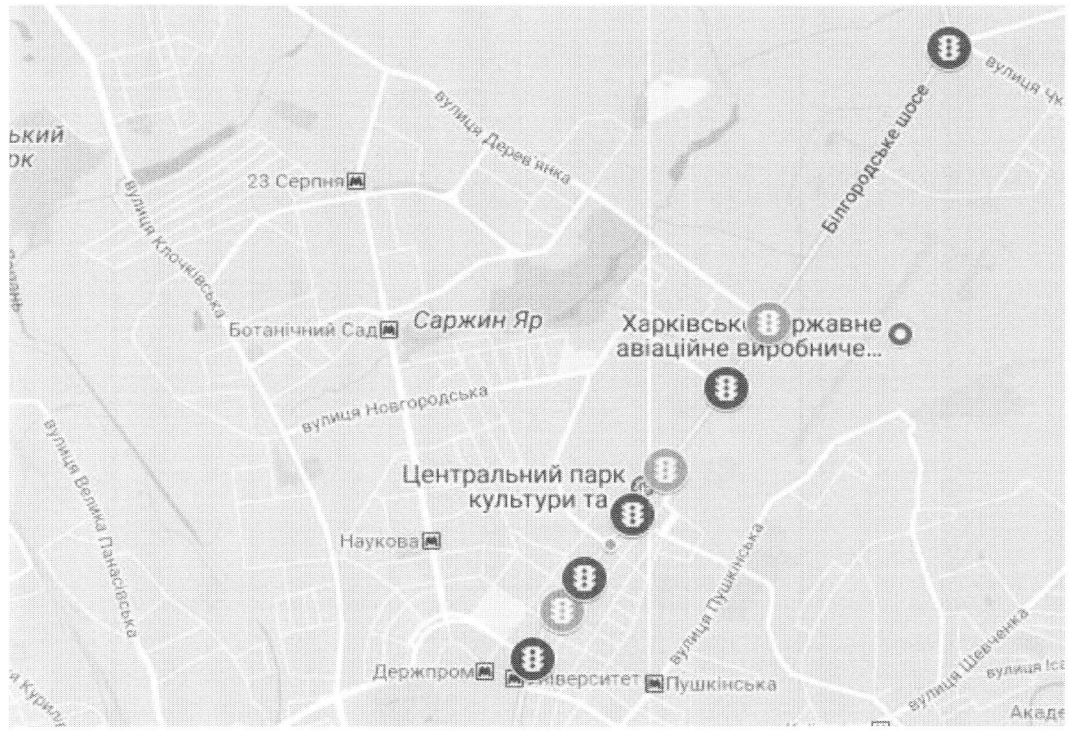

Figure 4 – The Arrangement of Traffic Lights within the Central Road

Traffic Light Group	Green Phase	Red Phase	Amber Phase	Green Phase
Blue	32(p1)	42(p2)	3	45(p3)
Orange	40(p4)	32(p5)	3	-
Green	42(p6)	35(p7)	3	-

Table 1 – Traffic Light Operating Time (s)

The AnyLogic software provides a unique suite of industry-specific tools in one package aimed to accomplish multimethod simulation modeling of business and industry infrastructures.

When simulation is performed with the AnyLogic (Personal Learning Edition for beginners and students), we can optimize only 7 parameters, therefore 9 traffic lights will be divided into 3 groups (Table 1).

The experiment will let us optimize the operation time of 9 traffic lights simultaneously. Fig. 5 shows a fragment of simulation modeling process.

At first, the simulation is accomplished without optimization. The results are shown in Fig. 6.

To carry out optimization, we set the minimum and maximum values for the traffic light operation and a time interval with which the parameters in the model will change.

As a result of optimization, we have obtained the best time for each parameter of the traffic light (Fig. 7). Then we will substitute these results into the developed model and carry out re-simulation (Fig. 8).

Once the optimization is carried out, the time of car passing through the traffic lights has decreased almost twice. Consequently, the traffic capacity of the road has increased.

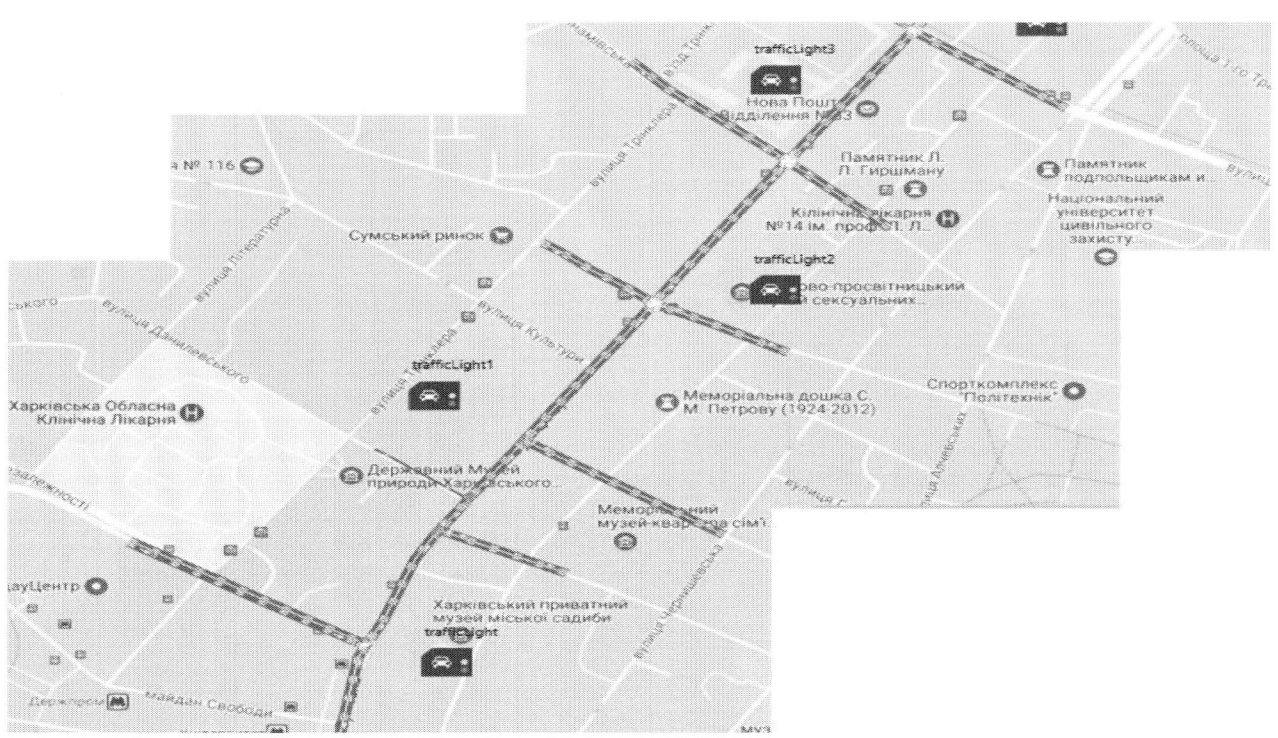

Figure 5 – A Fragment of Model Simulated with the AnyLogic Software

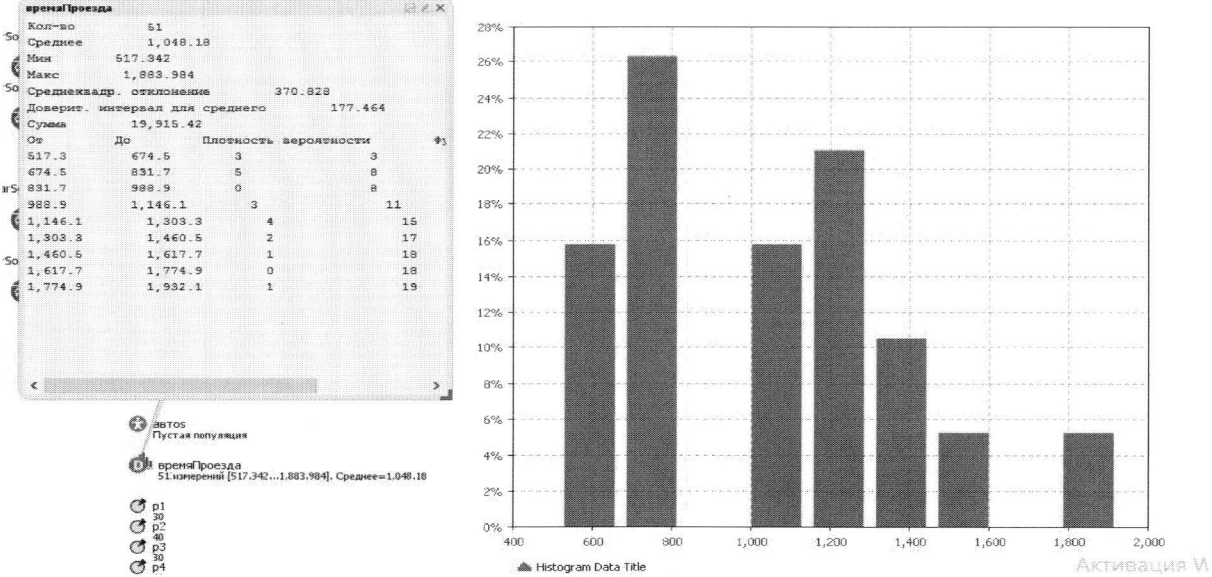

Figure 6 – The Results of Simulation prior to Optimization

движение2 : Optimization

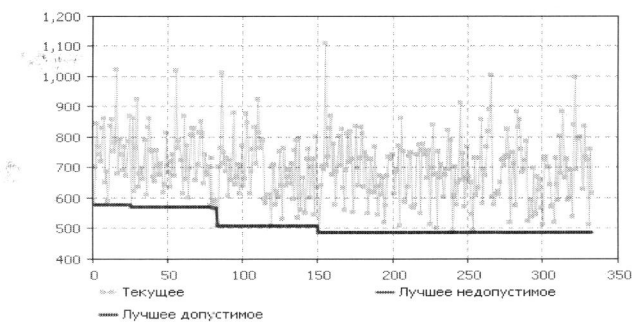

Figure 7 – The Results of Optimization with the AnyLogic Software

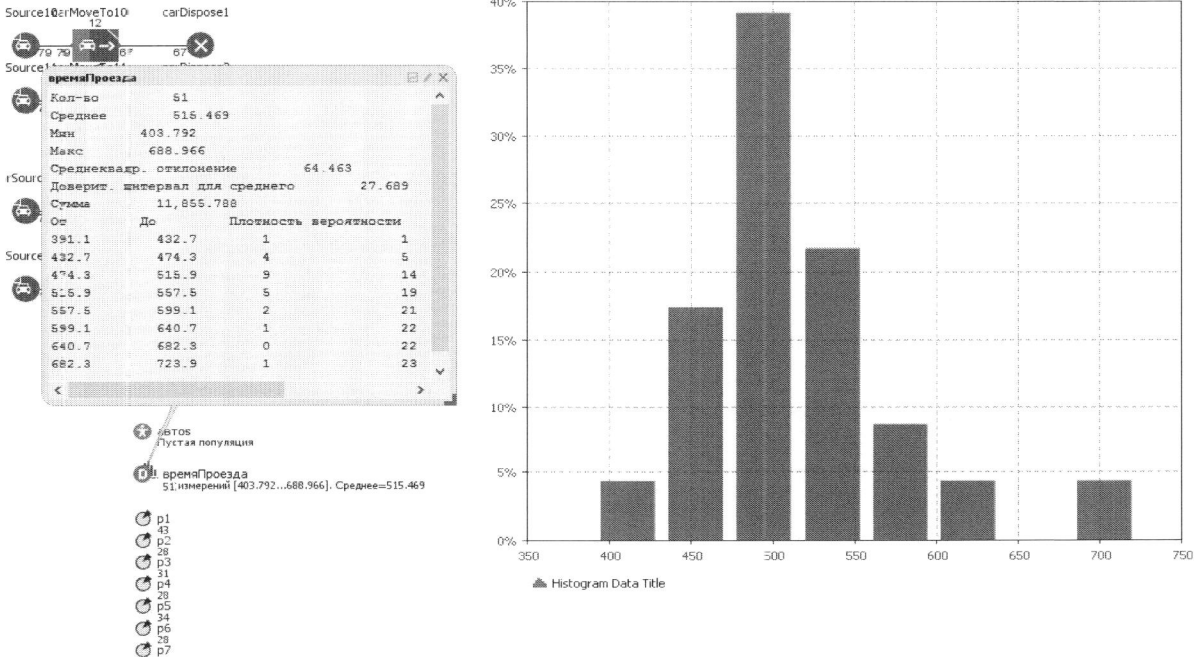

Figure 8 –The Results of Simulation after Optimization

Thus, the authors have made an attempt to simulate and optimize traffic capacity within the transportation infrastructure of the central part of the Kharkiv City as a case-study, using geo-informational technologies.

References

[1] Borisov, E.F. (2011) Gorodskaya infrastruktura i teoriya proektirovaniya dorozhnogo dvizheniya [City nfrastructure and Theory of Road Traffic]. Moscow: Vyisshee obrazovanie.

[2] Bulavina, L.N. (2009) Raschet propusknoy sposobnosti magistraley i uzlov [Calculations of Highway and Terminal Capacity].

[3] Mirotin, L.B. (1994) Transportnaia logistika [Transportation Logistics].Omsk.

[4] Mirotin, L.B. (2002) Transportnaia logistika [Transportation Logistics]. Moscow: Ekzamen.

[5] Kupriyashkin, A.G. (2015) Osnovy modelirovaniya sistem [Fundamentals of System Modeling]. Norilskiy industr. in-t. – Norilsk: NII.

[6] Katalevskiy, D.Yu. (2015) Osnovy imitatsionnogo modelirovaniya i sistemnogo analiza v upravlenii [Fundamentals of Simulation Modeling and System Analysis in Management]. Moscow: DeloPublishing House.

[7] Karpov, Yu. (2005) Imitatsionnoe modelirovanie sistem. Vvedenie v modelirovanie s AnyLogic 5 [Simulation Modeling System. Introduction into Modeling with AnyLogic 5]. Saint-Petersburg: BHV-Peterburg.

[8] Kiseleva, M.V. (2009) Imitatsionnoe modelirovanie sistem v srede AnyLogic [Simulation Modeling System in AnyLogic]. Yekaterinburg : UGTU – UPI.

WAREHOUSE COMPLEX DESIGN BASED ON FUZZY MODELING

Tetiana Pavlenko
Department of Economic Theory/Economics and Management Faculty
National Aerospace University "Kharkiv Aviation Institute", Ukraine

Olga Morozova
Department of Theoretical Mechanics, Mechanical Engineering and Robotic Systems/Engine Design Faculty
National Aerospace University "Kharkiv Aviation Institute", Ukraine

Kateryna Pechenizka
Department of Theoretical Mechanics, Mechanical Engineering and Robotic Systems/Engine Design Faculty
National Aerospace University "Kharkiv Aviation Institute", Ukraine

At present, any enterprise, regardless of its purpose and activities, deals with the warehouse economy. A modern warehouse complex is a technically sophisticated equipped facility that includes interconnected elements, has an appropriate structure and provides a number of services to change material flows, as well as to stock, process and distribute the goods between the consumers. In logistics warehouse complexes are exploited to modify and adjust inbound and outbound material (goods) flows that differ by their intensity and nature. Therefore, the main indicators of warehouse capacities directly depend on the nature of enterprise freight traffic [1].

Conditionally, the warehouse space can be divided into two parts, namely the main reserve area used for goods storage, and auxiliary areas used for other purposes. The reserve area is designed for basic technological operations, including storage of goods, expedition and processing. Auxiliary areas are designed for storing containers, placing engineering devices and utility lines, as well as for other required services. When drafting a warehouse, one needs to know the functions of each area, how to optimize their parameters and location, and determine the work efficiency.

Warehouse planning is to ensure effective arrangement and storage of goods, as well as the use of warehouse equipment and conditions for complete safety of goods. This makes up a principle for planning of inner warehouse areas that allows maintaining the continuity and uninterrupted flow of warehouse processes.

The total area of the warehouse comprises the main areas as follows [2]:

$$S_{total} = S_{cargo} + S_{aux} + S_{rec} + S_{o.p.} + S_{w.p.} + S_{put-away} + S_{retr.},$$

where S_{cargo} is a cargo pallet area used for the storage of goods (in pallets, piles and other ways of goods storing), $м^2$;

S_{aux} is an auxiliary (operational) area of aisles and lanes, $м^2$;

$S_{rec.}$ is an area for goods receiving, $м^2$;

$S_{o.p.}$ is an order-picking area, $м^2$;

$S_{w.p.}$ is an area of workplaces i.e. the area within the warehouse premises designed for the warehouse personnel, $м^2$;

$S_{put-away}$ is an area where the goods are prepared for put-away, $м^2$;

$S_{retr.}$ is an area where the goods are retrieved for shipping, $м^2$.

The main functional areas of the warehouse depend on the factors that cannot be always accurately estimated. Therefore, in this paper the authors offer to build an expert system based on fuzzy modeling. It will allow one to determine the warehouse areas according to the given criteria.

Thus, the purpose of this research is to create an expert system for the design of a warehouse complex using fuzzy simulation and design a warehouse for the storage of finished products, taking into account the expenses for its maintenance. The research focuses on a fuzzy modeling system as a tool used to calculate the warehouse complex areas.

The paper provides a definition and a full account of expert systems.

Expert systems are a class of computer programs that offer recommendations, conduct analyses, perform classifications and support decision-making processes [3]. They are applied for solving problems, whose solution requires an expert examination by a human being. Unlike programs that use procedural analysis, expert systems solve problems within a narrow domain (a specific area of examination) using logical assumptions. These systems can often find solutions to problems that are either ill-structured or inaccurately defined. Relying on heuristics, they compensate for the lack of structuring, which is useful in situations where an insufficient amount of required data or time makes the complete analysis impossible.

The expert systems employ the knowledge that is structured to simplify the decision-making process. A full-fledged expert system comprises four important components:

1) a knowledge base;

2) an inference engine;

3) a command interpreter;

4) an interface (explanation system).

To build the so-called fuzzy expert and/or control systems, one can use Fuzzy Logic Toolbox [4] as part of the MATLAB technical computing software. It is a package of applications for solving problems with fuzzy logic. Fuzzy inference systems created with the Fuzzy Logic Toolbox package can be integrated with the Simulink package tools. The latter allows one to simulate how all parts of the system behave. The simulation process clearly visualizes the results of mathematical modeling.

The key options of the Fuzzy Logic Toolbox are as follows:

• the construction of fuzzy inference systems (expert systems, fuzzy controllers, approximators of dependencies);

• the construction of adaptive fuzzy systems (fuzzy neural networks);

• interactive dynamic simulation in the Simulink.

Fuzzy inference systems created with the Fuzzy Logic Toolbox package can be integrated with the Simulink package tools. The latter allows one to simulate how all parts of the system behave. The simulation process clearly visualizes the results of mathematical modeling.

When building an expert system to determine the size of the warehouse premises, the following rules were set:

1) if the total area of the warehouse is determined, such constituents as cargo pallet area, auxiliary area, goods receiving area, area of workplaces and order-picking area are supposed to be determined as well;

2) if the cargo pallet area is determined, such constituents as the average stock level of goods and the averaged warehouse workload per one square meter of cargo pallet area of the warehouse are supposed to be determined as well;

3) the average stock level of goods depends on such parameters as the planned stock turnover, the number of days within the planned period and the number of units of goods;

4) if the auxiliary area is determined, the area of aisles and the parking area are supposed to be determined;

5) the area of aisles depends on the size of aisles for people and aisles for vehicles;

6) the parking area for vehicles depends on the number of vehicles and their dimensions;

7) the area of goods receiving depends on the amount of goods passing through this area and the approximate cost of storage within this area;

8) the area of workplaces depends on the number and the dimensions of workplaces;

9) the order-picking area depends on the following parameters: the total area of the warehouse and the coefficient that characterizes the dependence of the order-picking area on the cargo pallet area of the warehouse.

To generalize, all the factors characterizing the warehouse operation can be divided into four groups:

Group I. Factors that characterize the efficient use of the warehouse areas, namely storage capacity.

Group II Factors that characterize the efficiency of warehouse technological processes, namely cargo turnover, specific cargo turnover, uneven warehouse workload, service time for one order, etc.

Group III. Factors that characterize the level of cargo storage, including the number of cargo losses within a certain period of time, the ratio of losses per day to the total volume of goods stored per day, the amount of cargo returned due to damage.

Group IV. Economic factors that characterize the overall efficiency of the warehouse operation such as cost of storage, employee productivity and profit from warehousing operations.

In logistics, warehouse complex transforms input and output cargo (freight) flows that differ by their intensity and character. Therefore, the main factors of storage capacity will directly depend on the characteristics of cargo flows.

Warehouse zoning (differentiation between certain areas according to their function) is based on the processes inherent to a certain flow (flows). As a rule, the warehouse contains:

• an area for vehicle unloading (it can be located both inside and outside the warehouse premises);

• an area for goods receiving (including by the quantity and quality of goods);

• the main storage area;

• an order-picking area

• an area for goods expediting (dispatch and receipt of goods);

• an area for vehicle loading.

In addition, there is a reserve area provided for storing unscheduled consignments of goods sent for warehousing and an area for storing defective goods. Certainly, all major areas are to be interconnected by aisles and lanes whose size will be sufficient for people to walk and vehicles to transfer cargoes within the warehouse. The size is determined taking into account the dimensions of cargoes and forklifts, as well as the planned and / or actual cargo turnover. Let us proceed directly to the construction of a fuzzy expert system. The fuzzy expert system for calculating the total area of the warehouse has been created in the Simulink using the Fuzzy Logic Toolbox (Fig. 1).

The Simulink package is intended for simulating models that consist of graphic blocks whose properties (parameters) are specified. Model components are graphical blocks and models that are contained in a number of libraries. The models may contain various types of signal sources, virtual recording devices and animation graphics. By clicking the mouse twice on the model block, a window appears with a list of its parameters that can be modified by the user.

Simulation provides mathematical modeling of the constructed model with visual representation of the results.

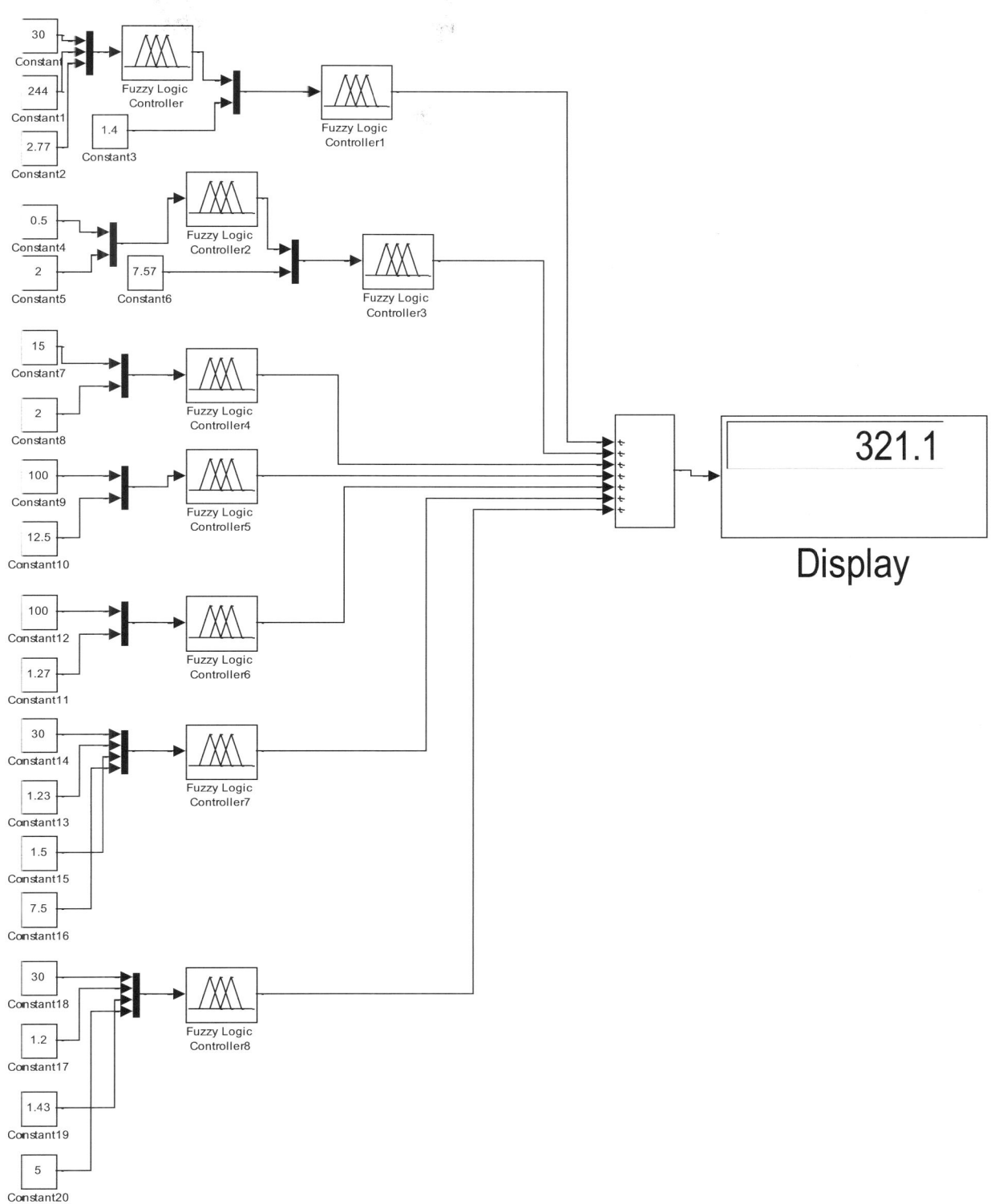

Figure 1 – Fuzzy Expert System created in the Simulink with the Fuzzy Logic Toolbox

Let us illustrate how a fuzzy expert system operates. The case study is the subsystem "Area for storing goods". Let us use the term set "Average Stock of Goods" represented by $T_1 =$ {"м", "с", "в"} as an input variable, where м is a small stock, с is an average stock, and в means a large stock, with the functions of the term membership (Fig. 2).

The term set "Averaged Warehouse Workload" will be used for an input value as an infinite set of $T_2 =$ {"м", "с", "в"} with the functions of the term membership (Fig. 3).

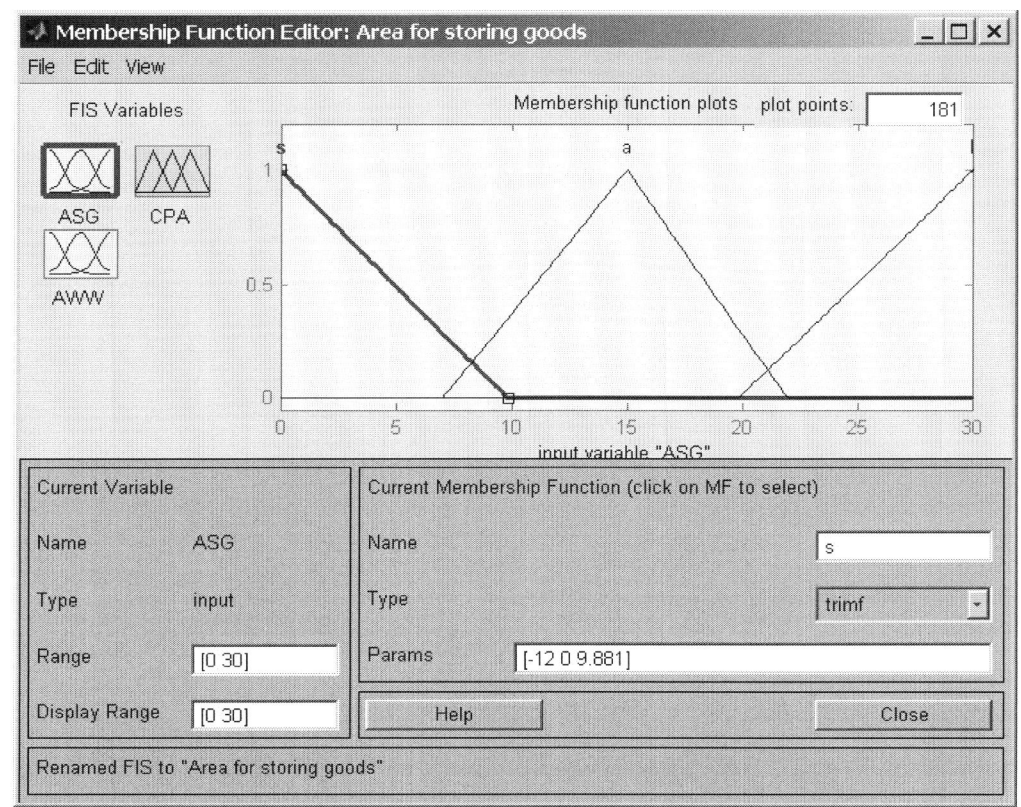

Figure 2 – Membership function "Average Stock of Goods"

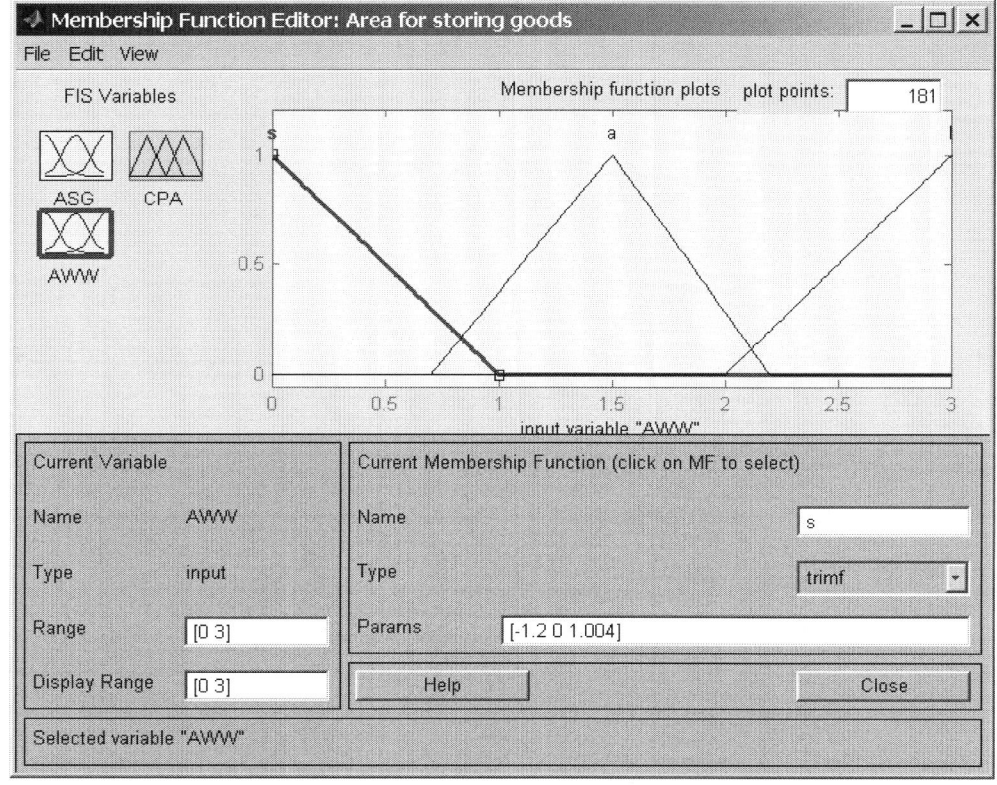

Figure 3 – Membership function "Averaged Warehouse Workload"

An infinite set of T_3 = {"м", "с", "в"} with the functions of the term membership (Fig. 4) will be used as another intermediate variable "Cargo Pallet Area". To build a set of purpose rules let us use nine rules of fuzzy products as in Fig.5. Let us consider the operation of "Cargo Pallet Area" subsystem (Fig. 6).

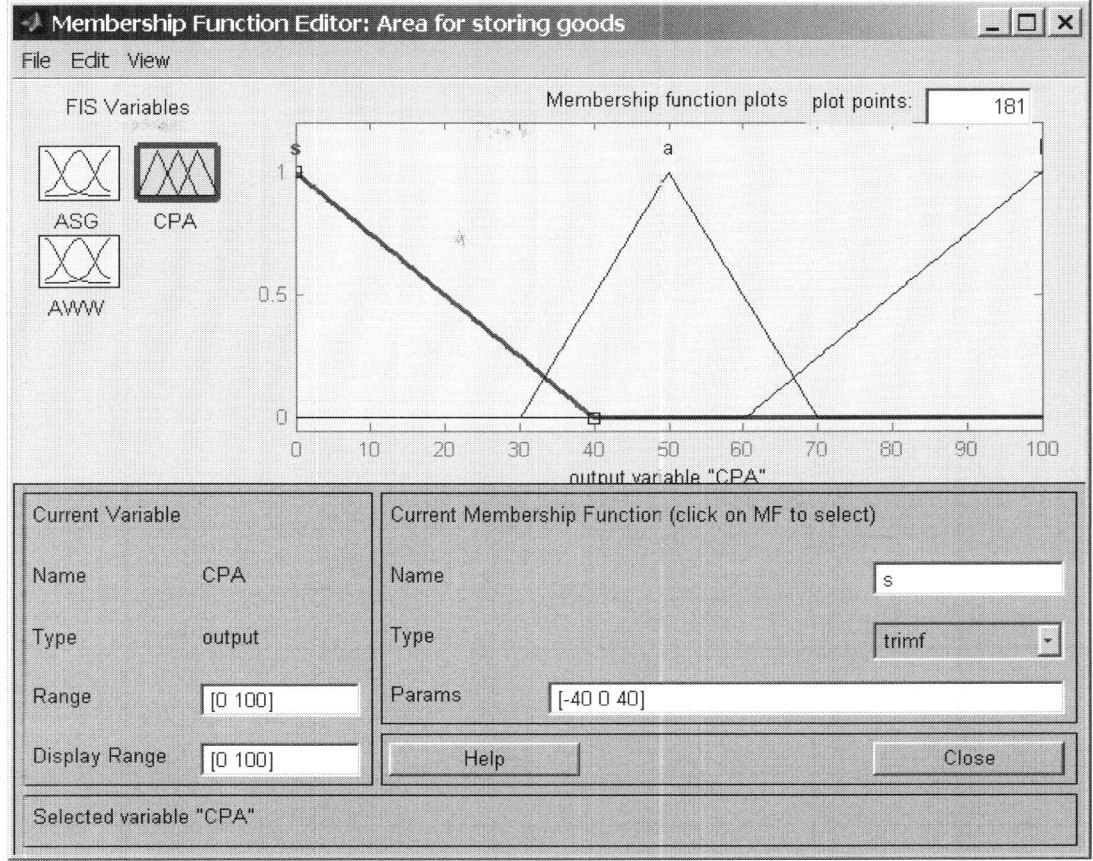

Figure 4 – Membership function "Cargo Pallet Area"

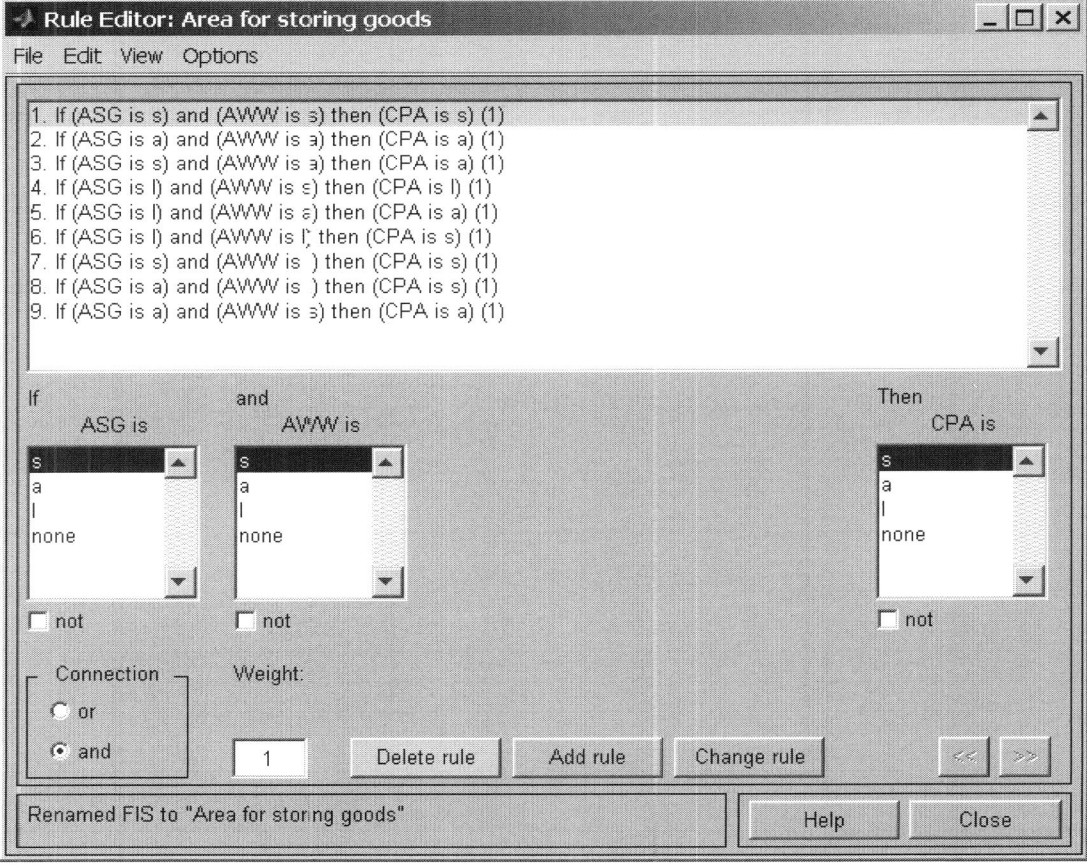

Figure 5 – Set of Rules for "Cargo Pallet Area"

135

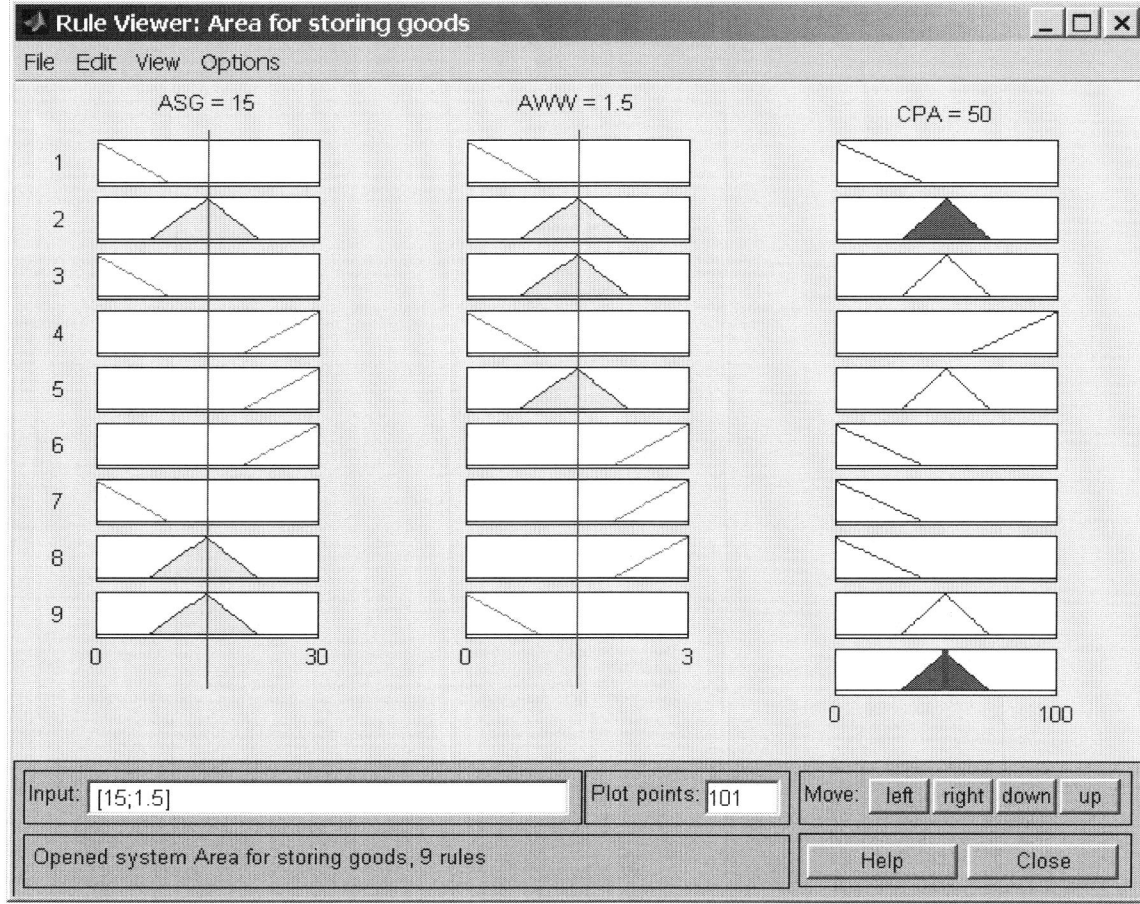

Figure 6 – Operation of "Cargo Pallet Area" subsystem

The same algorithm is applied for determining other areas within the warehouse. The Display block in Fig. 1 shows the total area of the warehouse complex.

To conclude, the authors of the present paper have built an expert system using the Fuzzy Logic Toolbox and Simulink packages in the MATLAB environment [5]. The expert system has been built as a fuzzy inference system for determining the areas within the warehouse complex, including its total area.

References

[1] Sergeev, V. I. (2008). Logistika: Informatsionnyie sistemyi i tehnologi [Logistics: Information Systems and Technologies. Moscow: Alfa-Press, 2008.

[2] Gadzhinskiy, A.M. (1999). Osnovyi logistiki [Fundamentals of Logistics]. Moscow: Inform.-vnedrench. tsentr "Marketing".

[3] Nerush, Yu.M. (2014). Proektirovanie logisticheskih sistem: uchebnik i praktikum dlya bakalavriata i magistraturyi [Logistics System Design for Students]. Moscow: Izd-vo Yurayt.

[4] Grigorev, M. N. (2015). Logistika [Logistics]. Mjscow: Izd-vo Yurayt.

[5] Rotshteyn, A.P. (1999). Intellektualnyie tehnologii identifikatsii: nechetkaya logika, geneticheskie algoritmyi, neyronnyie seti [Smart Technologies for Identification: Fuzzy Logic, Genetic Algoruthms and Neuron Networks]. Vinnitsa: UNIVERSUM-Vinnitsa.

[6] Zaychenko, Yu.P. (2004). Osnovi proektuvannya Intelektualnih sistem [Fundametals of Smart System Design]. Kyiv: Vidavnichiy Dlm «Slovo».

[7] Zimmermann H.-J. (1996). Fuzzy Set Theory and its Applications. 3rd ed. Dordrecht: Kluwer Academic Publishers.

[8] Leonenkov, A.V. (2005). Fuzzy modeling in MATLAB and fuzzyTech. St. Petersburg: BHV.

Committee of Reviewers

Univ.-Prof. Dr.-Ing. habil. Prof. E. h. Dr. h. c. mult. Michael Schenk
Institute of Logistics and Material Handling Systems
Otto von Guericke University Magdeburg, Germany
Fraunhofer Institute for Factory Operation and Automation IFF, Magdeburg, Germany

Prof. Dr. Béla Illés
Institute of Logistics
University of Miskolc, Hungary

Prof. Dr.-Ing. Dr. h. c. Norge I. Coello Machado
Department of Mechanical Engineering
Central University from Las Villas, Cuba

Dr.-Ing. Dr. h.c. (UCLV) Elke Glistau
Institute of Logistics and Material Handling Systems
Otto von Guericke University Magdeburg, Germany

PD Dr. rer. nat. habil. Juri Tolujew
Institute of Logistics and Material Handling Systems
Otto von Guericke University Magdeburg, Germany
Transport and Telecommunication Institute, Riga, Latvia

Prof. Dr.-Ing. Fabian Behrendt
Professor for Industrial Engineering
SRH Fernhochschule – The Mobile University, Riedlingen

LIST OF AUTHORS

Avdeikins, Aleksandrs
Transport and Telecommunication Institute, Riga, Latvia

Bányainé, Ágota Tóth, PhD
Institute of Logistics
University of Miskolc, Hungary

Bányai, Tamás; PhD
Institute of Logistics
University of Miskolc, Hungary

Béla, Illes; Prof. Dr.
Institute of Logistics
University of Miskolc, Hungary

Bezsmertna, Anastasiia
Department of Theoretical Mechanics and Engineering and Robotic Systems/ Aircraft Engines Faculty
National Aerospace University "Kharkiv Aviation Institute"

Borstell, Hagen; Dipl.-Sporting Dipl.-Ing.
Institute of Logistics and Material Handling Systems
Otto von Guericke University Magdeburg, Germany

Cabrera, Ernesto González, Ing.
Industrial Engineering Department, Central University from Las Villas, Cuba

Cao, Thanh Dung; M.Sc.
Institute of Logistics and Material Handling Systems
Otto von Guericke University Magdeburg, Germany

Castro, Roberto Cespón; DrC
Industrial Engineering Department, Central University from Las Villas, Cuba

Coello Machado, Norge Isaias; Prof. Dr.-Ing. Dr. h.c.
Mechanical Engineering Department, Central University from Las Villas, Cuba

Dvornikov, Mikhail
Department of Information Control Systems/ Aircraft Control Systems Faculty
National Aerospace University "Kharkiv Aviation Institute", Ukraine

Glistau, Elke; Dr.-Ing. Dr. h. c. (UCLV)
Institute of Logistics and Material Handling Systems
Otto von Guericke University Magdeburg, Germany

Goya Valdivia, Félix Abel; DrC
Chemical Engineering Department, Central University from Las Villas, Cuba

Illés, Béla; Prof. Dr.
Institute of Logistics
University of Miskolc, Hungary

Jackson, Ilya; PhD student
Department of Mathematical Methods and Modelling
Transport and Telecommunication Institute, Latvia

Juhász, János; PhD
Institute of Logistics
University of Miskolc, Hungary

Lang, Sebastian; M.Sc., M.Sc.
Institute of Logistics and Material Handling Systems
Otto von Guericke University Magdeburg, Germany

Malykhina, Iryna
Department of Theoretical Mechanics, Mechanical Engineering and Robotic Systems/Engine Design Faculty
National Aerospace University "Kharkiv Aviation Institute", Ukraine

Maure, Lissette Concepción; Ing.
Industrial Engineering Department, Central University from Las Villas, Cuba

Morozova, Olga
Department of Theoretical Mechanics, Mechanical Engineering and Robotic Systems/Engine Design Faculty
National Aerospace University "Kharkiv Aviation Institute", Ukraine

Nagy, Gábor, PhD student
Institute of Logistics
University of Miskolc, Hungary

Nefedkina, Olga
Department of Foreign languages/Humanities Faculty
National Aerospace University "Kharkiv Aviation Institute"

Neubert, Andreas; Dipl.-Inf.
Logistik Competence Center
PKE Deutschland GmbH, Germany

Pavlenko, Tetiana
Department of Economic Theory/Economics and Management Faculty
National Aerospace University "Kharkiv Aviation Institute", Ukraine

Pechenizka, Kateryna
Department of Theoretical
Mechanics, Mechanical
Engineering and Robotic
Systems/Engine Design Faculty
National Aerospace University
"Kharkiv Aviation Institute",
Ukraine

Polovynko, Volodymyr
Department of Theoretical
Mechanics, Mechanical
Engineering and Robotic
Systems/Engine Design Faculty
National Aerospace University
"Kharkiv Aviation Institute",
Ukraine

Pristupa, Larysa
Department of Theoretical
Mechanics, Mechanical
Engineering and Robotic
Systems/Engine Design Faculty
National Aerospace University
"Kharkiv Aviation Institute",
Ukraine

Rittberger, Sven; M.Sc.
Technology Centre for Production
and Logistics Systems (PULS)
Landshut University of Applied
Sciences, Germany

Rudenko, Nataliya
Department of Theoretical
Mechanics and Engineering and
Robotic Systems/ Aircraft Engines
Faculty
National Aerospace University
"Kharkiv Aviaion Institute"

Savrasovs, Mihails
Transport and Telecommunication
Institute, Riga, Latvia

**Schenk, Michael; Prof. Dr.-Ing.
habil. Prof. E. h. Dr. h. c. mult.**
Institute of Logistics and Material
Handling Systems
Otto von Guericke University
Magdeburg, Germany
Fraunhofer Institute for Factory
Operation and Automation IFF,
Magdeburg, Germany

Schneider, Markus; Prof. Dr.
Technology Centre for Production
and Logistics Systems (PULS)
Landshut University of Applied
Sciences, Germany

Simakovs, Andrejs
Trialto Latvia Ltd., Saulgozi,
"Dominante", Latvia

Solianyk, Tatiana
Department of Informatior Control
Systems/ Aircraft Control Systems
Faculty
National Aerospace University
"Kharkiv Aviation Institute",
Ukraine

Szentesi, Szabolcs, PhD student
Institute of Logistics
University of Miskolc, Hungary

Tamás, Péter, PhD
Institute of Logistics
University of Miskolc, Hungary

Thomas, Franziska; M.Sc.
Institute of Logistics and Material
Handling Systems
Otto von Guericke University
Magdeburg, Germany

**Tolujew, Juri, PD Dr. rer. nat.
habil.**
Institute of Logistics and Material
Handling Systems
Otto von Guericke University
Magdeburg, Germany
Transport and Telecommunication
Institute, Riga, Latvia

Trojahn, Sebastian, Dr.-Ing
Institute of Logistics and Material
Handling Systems
Otto von Guericke University
Magdeburg, Germany

Vasilyuk, Vadim
EOS Data Analytics, Ukraine

**Zadek, Hartmut; Univ.-Prof. Dr.-
Ing.**
Institute of Logistics and Material
Handling Systems
Otto von Guericke University
Magdeburg, Germany

IMPRESSUM

11th International Doctoral Students Workshop on Logistics
June 19, 2018 Magdeburg

Institut für Logistik und Materialflusstechnik
an der Otto-von-Guericke-Universität Magdeburg
Herausgeber:
Univ.-Prof. Dr.-Ing. habil. Prof. E. h. Dr. h. c. mult. Michael Schenk
Universitätsplatz 2 | 39106 Magdeburg
Telefon +49 391 6758710 | Telefax +49 391 6718074
michael.schenk@ovgu.de
http://www.ilm.ovgu.de

Redaktion: M.Sc. Niels Schmidtke
                Dr.-Ing. Dr. h.c. (UCLV) Elke Glistau
Titelbild: M.Sc. Niels Schmidtke
Foto Vorwort: Dirk Mahler
Bilder, Grafiken: Soweit nicht anders angegeben,
liegen alle Rechte bei den Autoren der einzelnen Beiträge.

Herstellung:
DocuPoint

Bibliografische Information der Deutschen
Nationalbibliothek:
Die Deutsche Nationalbibliothek verzeichnet diese Publikation in der
Deutschen Nationalbibliografie; detaillierte bibliografische Daten sind
im Internet über http://dnb.d-nb.de abrufbar.
ISBN 973-3-944722-71-9

© by Univ. Magdeburg, 2018
Univ. Magdeburg
Universitätsplatz 2 | 39106 Magdeburg
Telefon +49 391 67-52277 | Telefax +49 391 67-11153
sandra.scheer@ovgu.de
http://www.uni-magdeburg.de/ueberuns.html

Alle Rechte vorbehalten
Für den Inhalt der Vorträge zeichnen die Autoren verantwortlich.
Dieses Werk ist einschließlich aller seiner Teile urheberrechtlich
geschützt. Jede Verwertung, die über die engen Grenzen des
Urheberrechtsgesetzes hinausgeht, ist ohne schriftliche Zustimmung
des Verlages unzulässig und strafbar. Dies gilt insbesondere für
Vervielfältigungen, Übersetzungen, Mikroverfilmungen sowie die
Speicherung in elektronischen Systemen. Die Wiedergabe von
Warenbezeichnungen und Handelsnamen in diesem Buch berechtigt
nicht zu der Annahme, dass solche Bezeichnungen im Sinne der
Warenzeichen- und Markenschutz-Gesetzgebung als frei zu
betrachten wären und deshalb von jedermann benutzt werden
dürften. Soweit in diesem Werk direkt oder indirekt auf Gesetze,
Vorschriften oder Richtlinien (z.B. DIN, VDI) Bezug genommen oder
aus ihnen zitiert worden ist, kann der Verlag keine Gewähr für
Richtigkeit, Vollständigkeit oder Aktualität übernehmen.

© 06/2018 Institut für Logistik und Materialflusstechnik
an der Otto-von-Guericke-Universität Magdeburg